INFORMATION-PROCESSING CHANNELS IN THE TACTILE SENSORY SYSTEM

SCIENTIFIC PSYCHOLOGY SERIES

*Edited by Stephen W. Link, University of California, San Diego
and James T. Townsend, Indiana University*

INFORMATION-PROCESSING CHANNELS IN THE TACTILE SENSORY SYSTEM

A Psychophysical and Physiological Analysis

George A. Gescheider

Syracuse University, Institute for Sensory Research

John H. Wright

Ronald T. Verrillo (Deceased)

Syracuse University, Institute for Sensory Research

Psychology Press

Taylor & Francis Group

New York   London

Psychology Press
Taylor & Francis Group
711 Third Avenue,
New York, NY 10017

Psychology Press
Taylor & Francis Group
27 Church Road
Hove, East Sussex BN3 2FA

First issued in paperback 2015

Psychology Press is an imprint of the Taylor and Francis Group, an informa business

© 2009 by Taylor & Francis Group, LLC

International Standard Book Number-13: 978-1-138-88297-3 (pbk)
International Standard Book Number-13: 978-1-84169-896-0 (Hardcover)

Library of Congress Cataloging-in-Publication Data

Gescheider, George A.
    Information-processing channels in the tactile sensory system : a psychophysical
and physiological analysis / George A. Gescheider, John H. Wright, Ronald T. Verrillo.
        p. cm. -- (Scientific psychology series)
    Includes bibliographical references and index.
    ISBN 978-1-84169-896-0 (hardcover : alk. paper)
    1. Touch. I. Wright, John H., 1938- II. Verrillo, Ronald T. III. Title.

QP451.G47 2009
612.8'8--dc22   2008043780

Visit the Taylor & Francis Web site at
http://www.taylorandfrancis.com

and the Psychology Press Web site at
http://www.psypress.com

Contents

Preface

The purpose of this book is to address the question of whether information-processing channels operate in the sense of touch. Channels are known to operate in other sense modalities such as vision and audition. In these latter two senses, the receptors are localized within sensory organs, namely the eyes and ears. The function of the highly concentrated receptors in these senses is to collect a vast amount of information about the external world. Furthermore, the sensory processing of this information is facilitated by the existence of channels that serve to isolate specific features of the external environment. Information-processing channels, such as spatial frequency channels in vision and critical bands in hearing, therefore greatly enhance both the detection of and discriminability among external stimuli. In the tactile sensory system, the receptors are not highly localized in specialized organs, as they are in the visual and auditory systems, but instead are widely distributed throughout the skin of the entire body. However, they are not distributed in a uniform manner. In this respect, vision and audition are very different from touch. The fundamental question then is whether channels operate in the sense of touch as they do in vision and audition. A compelling argument is presented that channels do indeed operate in the highly sensitive glabrous skin of the hands and fingertips. However, it is unknown whether channels exist in other types of skin such as hairy skin and mucocutaneous skin.

Perhaps the earliest hint that channels operate in the sense of touch was when Békésy reported in 1939 that different neural systems appear to underlie the detection of low- and high-frequency vibratory stimuli. Further evidence for the existence of tactile channels emerged from work conducted at the Institute for Sensory Research at Syracuse University over a period of several decades. This work, in combination with the work of investigators in several other laboratories, constitutes the foundation for this book, which clearly shows that channels operate in the sense of touch and consequently play a major role in tactile perception.

Acknowledgments

Several individuals contributed to the ideas presented in this book. Although Ronald Verrillo contributed directly, he died prior to the work's completion. Stanley Bolanowski was to have been a co-author but died before we began working on it. We wish to acknowledge the contributions of Mark Hollins, who read early versions of the book and provided many valuable suggestions and much encouragement. His published work greatly facilitated the development of many ideas expressed here. We thank our colleague Burak Güçlü for his critical reading of this work. His insightful comments were invaluable. We also thank the series editors, Stephen Link and James Townsend, for their advice and encouragement and Lejanet Herrera for her editorial guidance throughout the production process. Lastly, we wish to acknowledge the contributions of Katheryne Gall, whose artwork made it possible to display complex ideas in an easily understandable graphic format.

1
Channels in Touch

Touch is hardly deserving of its characterization as a "minor sense." Mechanical stimulation of the skin produces a rich and complex variety of tactile sensations and perceptions. Consider, for example, how smooth a glass surface feels to the fingertip when it is rubbed across it, how rough sandpaper feels, how soft the fur of the family pet feels, and how the mechanical vibrations from nearby machinery feel. The sense of touch can also be used to process complex information, as the use of Braille by a visually impaired individual so amply attests. The sense of touch also combines with other senses whose receptors lie in the skin to produce more complex perceptions. For example, the perception of the wetness of a liquid results from tactile and thermal receptors acting in combination (Bentley, 1900).

A great deal has been learned about how the sense of touch operates by systematically varying the physical properties of tactile stimuli applied to the skin and comparing the psychophysical responses of the observer to the physiological responses of tactile nerve fibers. We believe that sufficient data now exist to make it possible to formulate a highly compelling and comprehensive theoretical model capable of integrating and explaining the diverse experimental findings obtained from this approach to the study of the tactile system. The operation of specific tactile neural systems as information-processing channels forms the basis for our theoretical model and, in addition, helps to explain the richness and variety of the tactile perceptions experienced in everyday life.

Information-processing channels in sensory systems are tuned to specific regions of the energy spectrum to which the sensory system responds; hence, channels function as *filters*. It is the relative activity levels in each of several channels that determine the qualitative and quantitative dimensions of perceptual experience.

The channel concept is not new to sensory science. Indeed, channels have been shown to operate in several sensory modalities, as seen in studies of critical bands in hearing (e.g., Fletcher, 1940; Greenwood, 1961; Scharf, 1961), color vision and luminance (e.g., DeValois & DeValois, 1975; Hurvich, 1981; Lennie & D'Zmura, 1988; Wald, 1964), visual contrast sensitivity (e.g., Campbell & Robson, 1968), taste (e.g., Henning, 1916b), smell (e.g., Henning, 1916a), and touch (e.g., Bolanowski, Gescheider, & Verrillo, 1994; Bolanowski, Gescheider, Verrillo, & Checkosky, 1988; Gescheider, Bolanowski, Pope, & Verrillo, 2002; Makous, Friedman, & Vierck, 1995; Verrillo & Gescheider, 1975).

A fundamental characteristic of a channel is its ability to process information to which it is tuned, independent of ongoing activity in other channels. For example, if a neural system operates as an independent channel, then stimuli presented to other tactile channels should have no effect on its sensitivity. Indeed, altering the sensitivity of a particular channel by adaptation or masking has no effect on the ability of other tactile channels to process information (Capraro, Verrillo, & Zwislocki, 1979; Gescheider, Bolanowski, & Verrillo, 2004; Gescheider, Frisina, & Verrillo, 1979; Gescheider, O'Malley, & Verrillo, 1983; Gescheider & Verrillo, 1979; Gescheider, Verrillo, & Van Doren, 1982; Hamer, Verrillo, & Zwislocki, 1983; Hollins, Goble, Whitsel, & Tommerdahl, 1990; Labs, Gescheider, Fay, & Lyons, 1978; Verrillo & Gescheider, 1977).

This book focuses on the information-processing channels that operate when tactile stimuli are applied to the glabrous skin of the hand. Although this glabrous skin constitutes only a small proportion of the total amount of a human's skin, it is the skin primarily responsible for the tactile perception of objects. The relatively high density of specialized mechanoreceptors in glabrous skin endows it with its exceptional capacity for processing the spatial and temporal properties of objects that contact it. Hence, it is not surprising that a vast amount of research has been done on the sensory properties of the glabrous skin of the hand, the results of which have greatly enhanced our understanding of the psychophysical and neurophysiological bases of tactile perception.

From these findings we shall show that the neural systems activated by the tactile stimulation of glabrous skin operate as independent channels. These channels independently process information in the early stages of tactile perception and combine their outputs at later stages within the central nervous system. It is much less certain that the neural systems mediating the processing of tactile stimulation of other types of skin such as hairy skin and mucocutaneous skin operate as channels. Indeed, it is possible that channels exist only for the glabrous skin of the hand. We shall show that channels serve to enhance the detection and discrimination of tactile stimulation, while enriching tactile experience through the blending of their neural outputs. These channel properties greatly enhance the perception of objects that contact the glabrous skin of the hand, and it is the hand that is primarily used to explore the spatial and temporal properties of objects. Because the skin of other parts of the body rarely performs this function, channels would be of little or no benefit there. Hence, it would not be surprising if channels were absent in hairy and mucocutaneous tissue. Although at least three neural systems mediate the detection of vibratory stimuli of varying frequency applied to the hairy skin of the forearm (Bolanowski et al., 1994), there is no evidence that they operate as independent channels. In fact, there is evidence that they do not (Hollins, Delemos, & Goble, 1991).

Historically, the study of tactile channels has progressed from the initial identification of separate neural systems, each responsive to specific types of stimuli, to the demonstration by psychophysical experimentation that these neural systems operate as independent channels. This book examines evidence supporting the hypothesis that tactile neural systems operate as information-processing channels in the sense of touch and considers the hypothesis that suprathreshold tactile perceptions result from interaction of the channels. Hence, the following problems are dealt with: (1) identifying specific neural systems responsible for mechanoreception, with special attention to their physiological and psychophysical characteristics; (2) testing the hypothesis that these specific mechanoreceptive neural systems function as information-processing channels; (3) presenting a multichannel model of mechanoreception to account for a wide variety of tactile sensory phenomena; and (4) specifying how the channels interact in the perception of suprathreshold stimuli capable of activating more than one channel. We turn first to the problem of identifying the individual neural systems that are optimally sensitive to specific aspects of tactile stimulation.

2

Identification of Specific Neural Systems Responsible for Mechanoreception

Introduction

As is true in other sensory systems, specific receptors in touch respond optimally to different types of stimulation. For example, Pacinian corpuscles, with their rapidly adapting nerve fibers, respond best to the rapid onset and offset of deformations of the skin and to high-frequency vibration; whereas Merkel cells, with their slowly adapting nerve fibers, respond optimally to steady deformations of the skin and to low-frequency vibration (Vallbo & Johansson, 1984). Inasmuch as the receptor selectively responds to a particular type of stimulus, the afferent nerve fiber from the receptor serves as a communication line to convey specific information to the brain about that particular type of stimulus. Indeed, the earliest psychophysical demonstration of the operation of specific tactile neural systems, each optimally responsive to different types of stimulation, came from two findings: thresholds for the detection of high-frequency vibration are lowest at 250–300 Hz and rapidly rise at lower and higher frequencies; and at low frequencies of vibratory stimulation, thresholds are nearly independent of stimulus frequency (Békésy, 1939; Gescheider, 1976; Talbot, Darien-Smith, Kornhuber, & Mountcastle, 1968; Verrillo, 1963; Verrillo, Fraioli, & Smith, 1969).

Shown in Figure 2.1 are threshold measurements for detecting vibration applied to the skin (Verrillo, 1963). In this experiment a circular contactor attached to a vibrator was placed in contact with the skin at the prominence on the palm at the base of the thumb, known as the thenar eminence. The contactor protruded up through a hole in the rigid surface upon which the observer's hand rested and was shaped to fit the contour of the skin. There was a 1-mm gap between the circularly shaped contactor and a rigid surrounding surface.

5

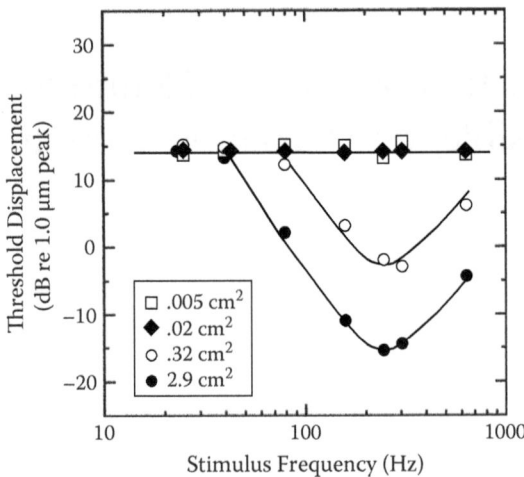

Figure 2.1 Detection thresholds as a function of stimulus frequency for stimuli applied to the thenar eminence through contactors of different sizes. Selected data from Verrillo (1963).

The rigid surface of the surround upon which the skin rested, by damping out traveling waves on the surface of the skin, served to control the area of stimulation by confining the vibratory stimulus to the area of the contactor. This was important because one of the objectives of the experiment was to investigate spatial summation by examining the effects on the detection threshold of changing the size of the contactor.

The data presented in Figure 2.1 were obtained with contactor sizes of 0.005 cm², 0.02 cm², 0.32 cm², and 2.9 cm². It can be seen that when the 2.9-cm² contactor was used, vibrotactile thresholds above 40 Hz were a U-shaped function of stimulus frequency. The greatest sensitivity occurred around 250 Hz, where the peak amplitude of vibration needed to exceed threshold was a value approximately -16 dB relative to 1 μm peak displacement of the skin. In contrast, below 40 Hz thresholds were much higher and completely independent of stimulus frequency.

In addition to illustrating the sensitivity of the tactile sensory system and its dependence upon the frequency of the vibratory stimulus, Verrillo's results also revealed that the effects of changing the size of the contactor depend strongly on stimulus frequency. When the size of the contactor was increased from 0.02 cm² to 0.32 cm² and then to 2.9 cm² (Figure 2.1), thresholds became progressively lower, provided the frequency of vibration was sufficiently high. This finding indicates that the sensitivity of the tactile system, as in other sensory systems, improves as the size of the stimulated area increases—a phenomenon known as *spatial summation*. However, for stimuli that were applied to the skin by contactors that were 0.02 cm^2 or smaller, no spatial summation was observed. Furthermore, the threshold-frequency functions for these small contactors are not U-shaped; their thresholds are uniformly high at all stimulus frequencies. Finally, below 40 Hz, the detection threshold is not affected by changing the size of the contactor, which shows that there is no spatial summation at low-stimulus frequencies. Because understanding the phenomenon of spatial summation is central to understanding the concept of tactile channels, the mechanisms underlying spatial summation will be discussed in great detail later in this book.

Duplex Model of Mechanoreception

From these findings, Verrillo (1963) concluded that there are at least two receptor systems responsible for the detection of mechanical disturbances of the skin. One system was referred to as the Pacinian (P) system because its receptors are Pacinian corpuscles (Verrillo, 1966). This system is capable of spatial summation and is responsible for the U-shaped frequency selectivity observed at relatively high frequencies when all but the smallest contactors are used. The other system was referred to as the non-Pacinian (NP) system because its receptors are other types of receptors in the skin. This system is not capable of spatial summation and its sensitivity is independent of stimulus frequency, which explains the flat frequency-sensitivity function observed at all frequencies with small contactors and at low frequencies with large as well as small contactors. These results, and their strong implication that two independent receptor systems mediate the detection of tactile stimuli, became the foundation of Verrillo's duplex model of mechanoreception (Verrillo, 1968). This model became the basis for much of the theoretical work that followed on the nature of tactile channels and must be discussed in detail.

Central to the duplex model is a fundamental principle of sensory processing, which states that the psychophysical detection threshold is always determined by the neural system with the lowest activation threshold (see Gescheider, 1997). This principle is illustrated in Figure 2.2. When the size of the contactor is 0.005 or 0.02 cm^2 (Figure 2.2A and 2.2B), the stimulus is always detected by the NP system, with its flat frequency-sensitivity function and absence of spatial summation. Thus, changing the frequency of the stimulus or the size of the skin area stimulated has no effect on the detection threshold. With very small contactors the threshold of the P system remains above that of the NP system at all stimulus frequencies, and, consequently, the P system is never activated during stimulus detection by the NP system. Only when the contactor is larger does the P system play a role in stimulus detection.

When the size of the contactor is 0.32 cm^2 (Figure 2.2C), the threshold for detecting high-frequency stimuli above 80 Hz is mediated by the P system

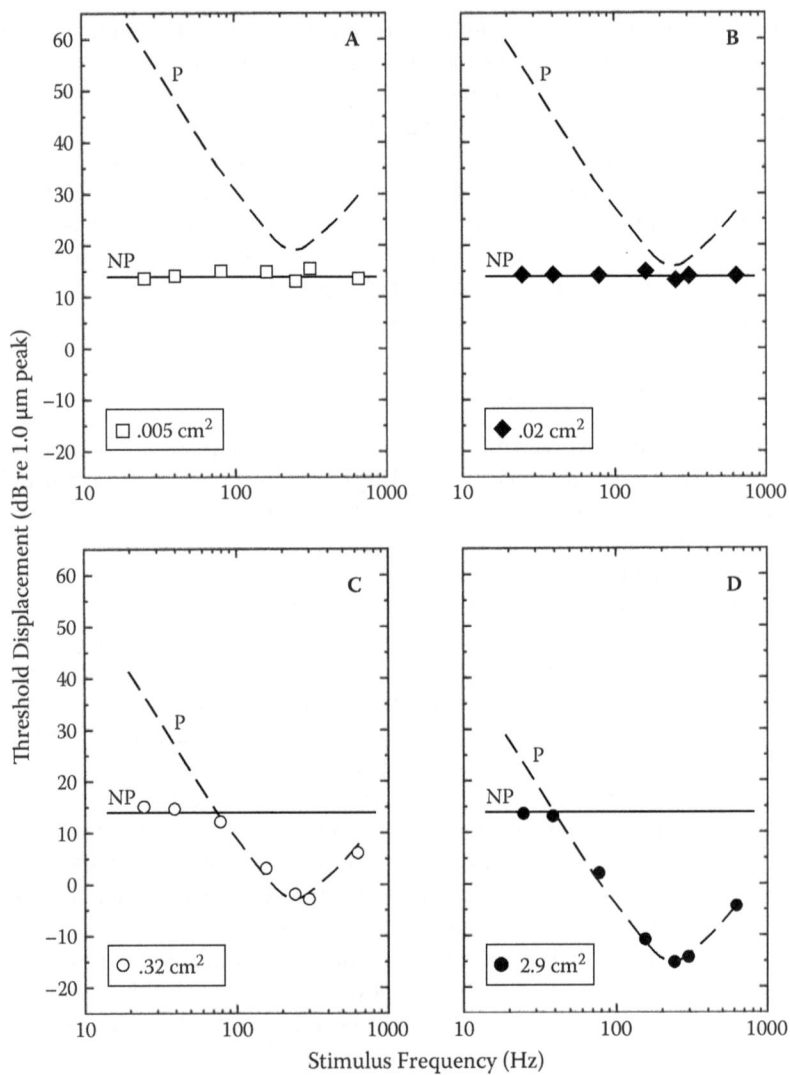

Figure 2.2   Data of Figure 2.1 replotted on a separate graph for each contactor size. The frequency-selectivity functions of the theoretical P system capable of spatial summation and the NP system not capable of spatial summation account for the threshold measurements made with each contactor. Selected data from Verrillo (1963).

because at these frequencies its activation threshold is lower than that of the NP system. Below 80 Hz the NP system, with its flat-frequency characteristic, determines the detection threshold because at these lower frequencies the activation thresholds of the P system have become higher than those of the NP system. When the size of the contactor is further increased to 2.9 cm² (Figure 2.2D), the activation threshold of the P system becomes even lower as a result of spatial summation, and the P system now determines threshold at all stimulus frequencies above 40 Hz. Below 40 Hz the NP system determines threshold because at low frequencies its threshold remains below that of the P system.

In the analysis presented above, a constant slope of approximately -12 dB per doubling of stimulus frequency is proposed for the sensitivity function of the P system at frequencies between 15 Hz and 150 Hz; but in the data presented thus far, this slope is evident only between 40 Hz and 150 Hz. This is seen in the detection thresholds obtained with the 2.9-cm² contactor (Figure 2.2D). The rationale for extrapolating the P system's slope of -12 dB per doubling of frequency down to 15 Hz is derived from an experiment in which the NP system was selectively adapted to raise its threshold function above that of the P system at all stimulus frequencies (Verrillo & Gescheider, 1977).

As seen in Figure 2.3, adaptation with a 10-Hz stimulus presented for 10 minutes at an intensity of 30 dB above the detection threshold produced a threshold function with a constant slope of approximately -12 dB per doubling of stimulus frequency down to frequencies as low as 15 Hz. Furthermore, the finding that psychophysical thresholds for the detection of low- but not high-frequency stimuli are elevated by a low-frequency adapting stimulus provides further support for the hypothesis that the P and NP systems are separate and independent. This suggests that they operate as independent channels, each unaffected by adaptation of the other. When a stimulus is detected by the P system, the sensitivity of this system is not affected by the 10-Hz adapting stimulus, provided the intensity of the adapting stimulus is below that needed to activate the P system. In contrast, the 10-Hz adapting stimulus has the effect of elevating the thresholds of the NP system, with its flat-frequency characteristic, to a point where the thresholds of the P system become lower than those of the adapted NP system at all stimulus frequencies. Consequently, all stimuli are now detected at threshold by the P system, and its frequency characteristic of -12 dB per doubling of frequency is revealed across the entire range of stimulus frequencies. This is because the adaptation of the NP system has no effect on the sensitivity of the P system.

Further evidence supporting the conclusion that the P and NP systems are independent with respect to adaptation is the complementary finding that adaptation with a high-frequency vibratory stimulus elevates the thresholds for detecting high- but not low-frequency stimuli (Gescheider, Capraro, Frisina, Hamer, & Verrillo, 1978). When the frequency of the adapting stimulus is optimally suited for P-system stimulation, namely 250 Hz, the threshold function for the P system, with its characteristic -12 dB per doubling of stimulus frequency, is raised by a constant amount across all frequencies of the test stimulus (Figure 2.4). At lower frequencies the thresholds of the NP system, with its characteristic flat frequency function, are not affected by a 250-Hz

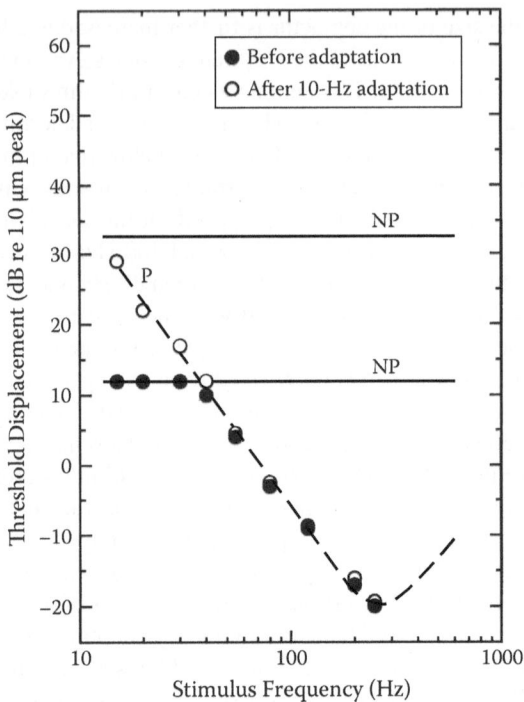

Figure 2.3 Detection thresholds as a function of stimulus frequency for stimuli applied to the thenar eminence before and after adaptation for 10 minutes with a 10-Hz stimulus applied to the test site at an intensity of 30 dB above the unadapted 10-Hz threshold. The dashed line represents the threshold function of the P system, and the solid lines represent the threshold function of the NP system before (lower) and after (upper) adaptation. Data from Verrillo and Gescheider (1977).

adapting stimulus because this stimulus activates the P but not the NP system. It will become apparent in the discussion of tactile channels that the discovery that adaptation occurs within but not across systems provides compelling evidence in support of the multichannel model presented later in this book.

Spatial and Temporal Summation in the P System

The dependence of spatial summation in the tactile system on stimulation of Pacinian corpuscles is illustrated in Figure 2.5, where detection thresholds are plotted as a function of the size of the contactor. Only when high-frequency stimuli capable of exciting Pacinian corpuscles are applied to an area of the skin containing these receptors (250-Hz stimulus applied to the thenar eminence) does the threshold decrease as the size of the contactor increases. In

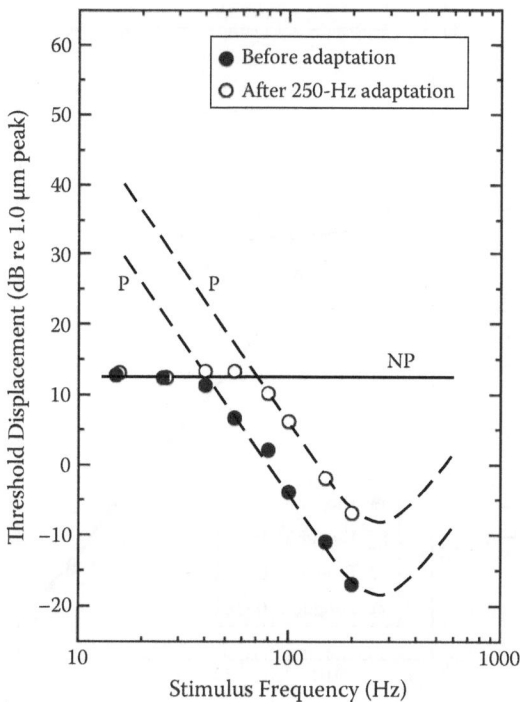

Figure 2.4 Detection thresholds as a function of stimulus frequency for stimuli applied to the thenar eminence before and after adaptation for 10 minutes with a 250-Hz stimulus applied to the test site at an intensity of 30 dB above the unadapted 250-Hz threshold. The solid line represents the threshold function of the NP system and the dashed lines represent the threshold function of the P system before (lower) and after (upper) adaptation. Data from Gescheider et al. (1979).

contrast to the spatial summation exhibited by the P system is its absence when the NP system detects the stimulus. It can be seen that the size of the contactor has no effect on the detection threshold when the stimulus is applied to the dorsal surface of the tongue, which contains no Pacinian corpuscles, or when it is applied to the thenar eminence at frequencies (1 Hz or 25 Hz) too low to excite Pacinian corpuscles.

In addition to the high-frequency tuning of the P system and its capacity for spatial summation is its unique capacity for temporal summation (Gescheider, 1976; Gescheider, Beiles, Bolanowski, Checkosky, & Verrillo, 1994; Gescheider, Berryhill, Verrillo, & Bolanowski, 1999; Gescheider & Joelson, 1983; Verrillo, 1965). As seen in Figure 2.6, when the P system is isolated by applying a high-frequency 250-Hz stimulus through a large 3-cm² contactor to the thenar eminence containing many Pacinian corpuscles, the detection

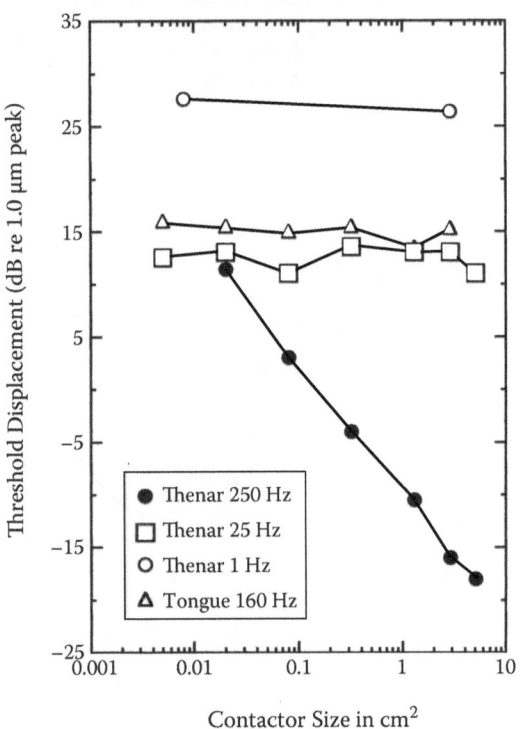

Figure 2.5   Detection thresholds as a function of contactor size for 25-Hz and 250-Hz stimuli applied to the thenar eminence (Verrillo, 1963), for 1-Hz stimuli applied to the thenar eminence (Bolanowski et al., 1988), and for 160-Hz stimuli applied to the tongue (Verrillo, 1968).

threshold decreases as stimulus duration increases, up to a duration of about 1 second. However, when the same stimulus is applied through a small 0.01-cm^2 contactor incapable of activating the P system but capable of activating NP receptors, the detection threshold becomes independent of stimulus duration. Also seen in Figure 2.6 is the finding that temporal summation is absent when a low-frequency 30-Hz stimulus that excites only NP receptors is presented to the thenar eminence or when a high-frequency 250-Hz stimulus is presented to the dorsal surface of the tongue—an area known to be lacking in Pacinian corpuscles (Verrillo, 1968).

These results clearly indicate that the P system is capable of temporal summation, whereas the NP system is not. Furthermore, temporal summation in the P system can be predicted from Zwislocki's theory, in which it is assumed that neural activity increases as time of exposure to the stimulus increases (Zwislocki, 1960). A close correspondence is seen between the

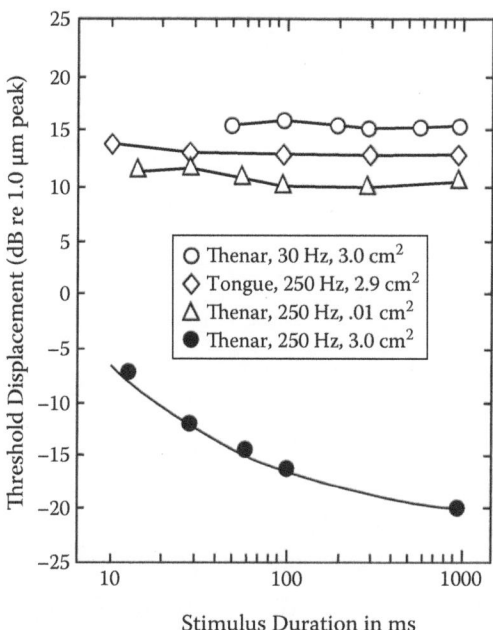

Figure 2.6 Detection thresholds as a function of stimulus duration for 250-Hz stimuli applied to the thenar eminence through a 0.01-cm² and through a 3-cm² contactor, for 30-Hz stimuli applied to the thenar eminence through a 3-cm² contactor (Gescheider, 1976), and for 250-Hz stimuli applied to the tongue through a 2.9-cm² contactor (Verrillo, 1968). The curve represents the change in threshold as a function of stimulus duration predicted from Zwislocki's theory of temporal summation (Zwislocki, 1960).

predicted decrease in threshold as stimulus duration increases (solid curve) and the data obtained for the P system with 250-Hz stimuli applied through the 3-cm² contactor. Thus, we can state with confidence that both temporal and spatial summation are exclusive properties of the P system.

Figure 2.6.

3

The Neural Bases of the Tactile Systems

Introduction

It is clear that at least two neural systems underlie the detection of vibrotactile stimuli—the highly tuned P system, with its U-shaped frequency characteristic and its capacity for spatial and temporal summation, and the NP system, with its flat frequency characteristic and absence of spatial and temporal summation. But what are the specific physiological mechanisms underlying these systems? To answer this question, the first thing that must be considered is the properties of the tactile mechanoreceptors and their associated nerve fibers in the glabrous skin of the hand, where there are not just two but at least four known types of mechanoreceptors and associated nerve fibers that mediate the sense of touch.

Anatomy of Tactile Receptors

The largest and by far the most investigated tactile receptor is the *Pacinian corpuscle*. Its afferent nerve fiber is located at the center of a capsule consisting of nearly 120 alternating layers (*lamellae*) of tissue and fluid. It is oval shaped and in cross-section resembles an onion. On average it is about 1 mm in length and is located deep in the dermis near major nerve trunks and arteries (see Figure 3.1). Each Pacinian corpuscle is innervated by a single nerve fiber that extends from the capsule to the lower brain stem.

A second tactile receptor in glabrous skin is the *Meissner corpuscle*. Located in the superficial region of the dermis, each Meissner corpuscle is tucked into a dermal projection near the epidermis (*dermal papilla*). The entire capsule hangs hammocklike by connective tissue from the walls of the dermal papilla. A single nerve fiber innervates one or more Meissner corpuscles.

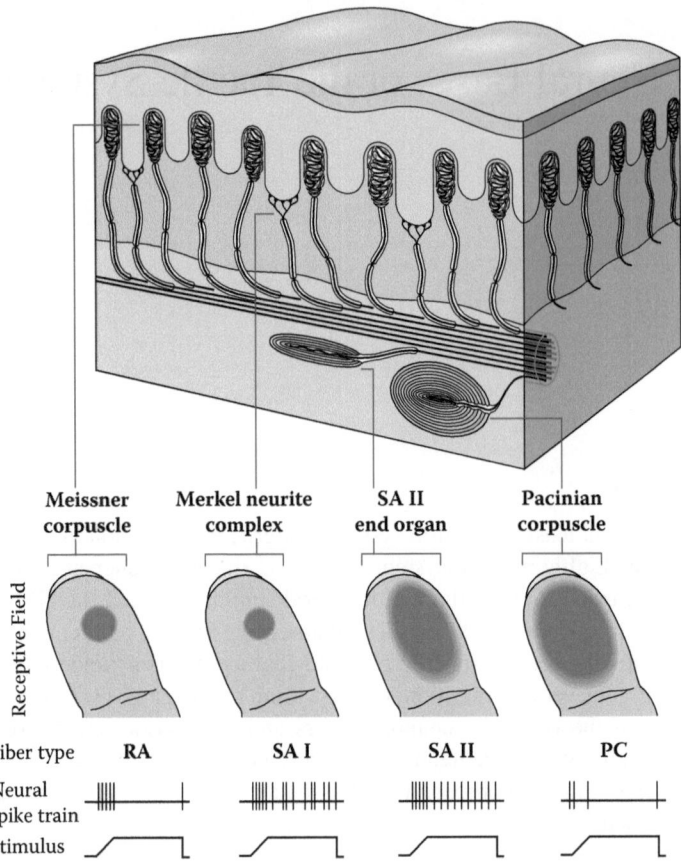

Figure 3.1 Tactile receptors and nerve fibers in glabrous skin.

A third tactile receptor is the *Merkel-neurite complex*. Merkel neurites have flat, leaf-like structures arrayed linearly along the lower aspect of the intermediate ridges of the epidermal extensions into the dermis. Multiple nerve fibers course through these leaf-like structures from one to another, and a single fiber may innervate many receptors.

It has traditionally been thought that the fourth major tactile receptor in glabrous skin is the *Ruffini ending*. However, in recent anatomical work, the classic Ruffini ending described by Chambers, Andres, von Düering, and Iggo (1972) has not been found in glabrous skin. It is the considered opinion of a growing number of anatomists that Ruffini-end organs are found primarily and probably exclusively in hairy skin (Dellon, 1981; Paré, Behets, & Cornu, 2003; Paré, Smith, & Rice, 2002). Thus, the tactile receptors found in glabrous skin that were formerly thought to be Ruffini endings appear to be an entirely different structure, possibly *Paciniform endings*, which are much smaller and simpler in structure than Pacinian corpuscles.

Neurophysiology of Tactile Receptors and Their Nerve Fibers

The mechanoreceptors can be classified in terms of the unique response properties of their nerve fibers. The discovery of these fiber types came largely from the electrophysiological recording of the activity from single peripheral nerve fibers in awake humans (Torebjörk & Ochoa, 1980; Vallbo, 1981; Vallbo & Johansson, 1976, 1984) and also from the earlier work on the response characteristics of tactile peripheral fibers in subhuman animals (Chambers et al., 1972; Iggo, 1963; Iggo & Muir, 1969; Iggo & Ogawa, 1977; Jänig, Schmidt, & Zimmermann, 1968; Lindblom, 1965; Lindblom & Lund, 1966). In the studies with humans, the skin was depressed with a probe, and the pattern of action potentials from a single nerve fiber produced by varying the position of the probe was recorded from a microelectrode implanted through the skin near the nerve fiber. In this way it was possible to define the fiber's receptive field as an area on the skin within which depressing the skin changed the neural activity of the fiber. Because receptive field size was very different for different fiber types, this became a useful way to classify them.

For example, Pacinian fibers have very large receptive fields, whereas the receptive fields of Meissner and Merkel fibers are much smaller (Figure 3.1). Fiber types can also be distinguished by the rate at which their neural response to the stimulus adapts. Adaptation rate is defined as the rate at which the frequency of action potentials diminishes over time when the stimulus probe is pressed into the skin and held at a constant depth within the receptive field of the nerve fiber. The classes of fibers identified in this way were: (1) rapidly adapting Pacinian-corpuscle (PC) fibers—these fibers have large receptive fields; (2) rapidly adapting (RA) fibers associated with Meissner corpuscles—these fibers have small receptive fields; (3) slowly adapting (SA I) fibers associated with Merkel-neurite complexes—these fibers have small receptive fields; and (4) slowly adapting (SA II) fibers associated with SA II end organs—these fibers have large receptive fields (see Figure 3.1 and Table 3.1). Because the end organ for the SA II nerve fiber—previously thought to be the Ruffini ending—is currently unknown, although thought to be the Paciniform ending, it is henceforth designated in this monograph as the SA II end organ.

The rapidly adapting fibers respond strongly when a probe is depressed into the skin within the fiber's receptive field and sometimes when withdrawn, but not during sustained stimulation when the skin is continuously displaced by the probe by a constant amount. In contrast, slowly adapting fibers respond as long as the probe indents the skin, although at a reduced level following an

Table 3.1 Tactile Mechanoreceptors and Associated Nerve Fibers in Glabrous Skin

		Receptive Field Size	
		Small	Large
Slow (Adaptation Rate)		Merkel neurite complexes with SA I fibers	SA II end organs with SA II fibers
Fast (Adaptation Rate)		Meissner corpuscles with RA fibers	Pacinian corpuscles with PC fibers

initial relatively intense response at the onset of stimulation (see neural spike train in Figure 3.1).

Neural Bases of the P and NP Systems

What is the relationship among these four nerve-fiber types with their specific receptors and the P and NP systems? Verrillo (1966) definitively answered this question for the P system by identifying the Pacinian corpuscle as the receptor responsible for the U-shaped portion of the frequency-response curve for detection thresholds. When the intensities of stimulation needed to produce either a detection response in a human observer or a criterion amount of neural activity in the excised Pacinian corpuscles of a cat (Sato, 1961) are plotted as a function of stimulus frequency, there is a striking correspondence between the psychophysical and neurophysiological results (Figure 3.2). The U-shaped portions of the psychophysical threshold-frequency function and the threshold function of the Pacinian corpuscle are essentially identical.

The detection of relatively high-frequency vibration by the P system mediated by the neural responses of Pacinian corpuscles was subsequently confirmed in the work of Mountcastle and his associates (LaMotte & Mountcastle, 1975; Mountcastle, LaMotte, & Carli, 1972; Mountcastle, Talbot, & Kornhuber, 1966; Talbot et al., 1968). In this work the psychophysical responses of humans and monkeys were compared with the neurophysiological responses of single peripheral mechanoreceptive nerve fibers in the monkey. The average neurophysiological threshold functions for PC and RA fibers from the work of Talbot et al. (1968) are shown in Figure 3.3. These investigators found that the function describing the neural threshold of PC fibers as a function of stimulus frequency parallels the high-frequency portion of the psychophysical threshold-frequency function, whereas at low frequencies the threshold responses of RA fibers parallel the relatively flat function describing how psychophysical

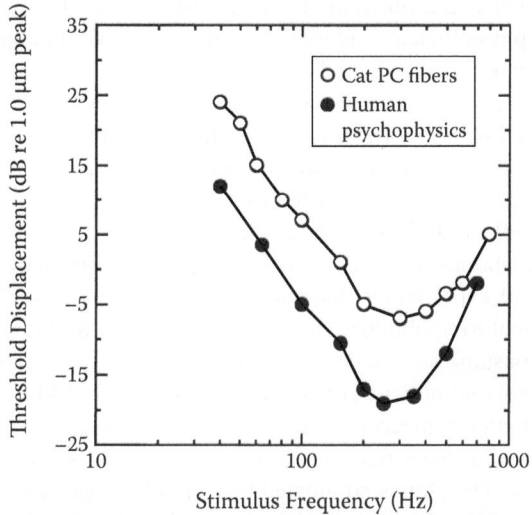

Figure 3.2   Threshold displacement as a function of stimulus frequency for the psychophysical thresholds of the P system measured on the thenar eminence of a human and for the neural thresholds of Pacinian corpuscles of a cat (Verrillo 1966).

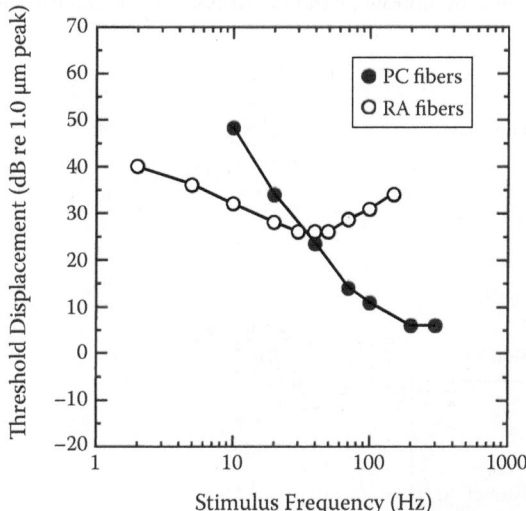

Figure 3.3   Threshold displacement as a function of stimulus frequency for monkey PC and RA fibers. From Talbot et al. (1968).

thresholds change as a function of stimulus frequency. These findings provide strong neurophysiological evidence in support of Verrillo's duplex model.

Additionally, in a study by Bolanowski and Verrillo (1982), the effects on detection thresholds of skin temperature at the surface of the observer's hand were compared with the effects on neural-activation thresholds of the temperature of the bathing solution in which excised Pacinian corpuscles from cats were immersed. The apparatus used to measure psychophysical thresholds in this study is illustrated in Figure 3.4 and represents the technology used in many tactile laboratories today to maintain precise control of the stimulus in psychophysical experiments. Vibratory stimuli are produced by applying an electrical signal to a vibrator. The contactor attached to the vibrator is maintained at a constant static indentation of approximately 0.5 mm into the skin, and the vibratory stimuli are applied relative to this value. The amplitude of the vibratory stimulus is measured either by an accelerometer or a linear variable displacement transducer that senses the displacement of the moving element of the vibrator. The skin is stimulated by a circular contactor, contoured to fit the curvature of the skin, attached to the moving element of the vibrator, and separated from the rigid surround by a 1-mm gap. The rigid surround confines the stimulus to the immediate area of the contactor by damping the spread of surface waves of vibration on the skin beyond the contactor. Skin temperature is held constant by using a device that circulates water of the appropriate temperature through the hollow chambers of both the contactor and the surround.

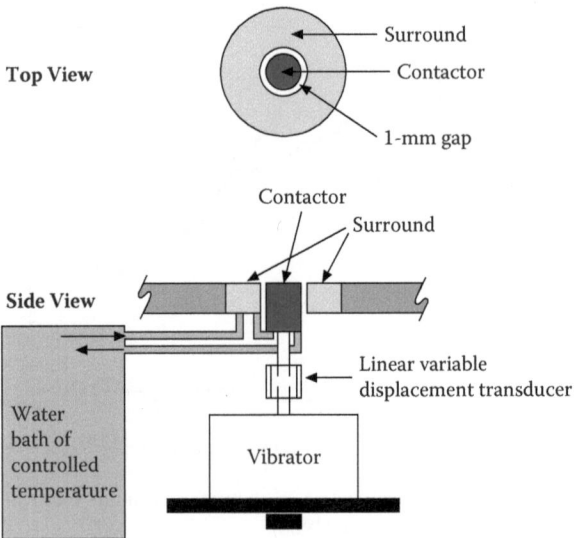

Figure 3.4 Illustration of the apparatus used by Bolanowski and Verrillo (1982) to measure psychophysical thresholds.

The results (Figure 3.5) show that variation in temperature had a remarkably similar effect on the psychophysical and neural-activation thresholds. In both measures, maximum sensitivity shifted to higher frequencies as temperature increased and thresholds tended to decrease. The high correlation between the neural-activation thresholds of PC fibers and the psychophysical thresholds as temperature and stimulus frequency change constitutes compel-

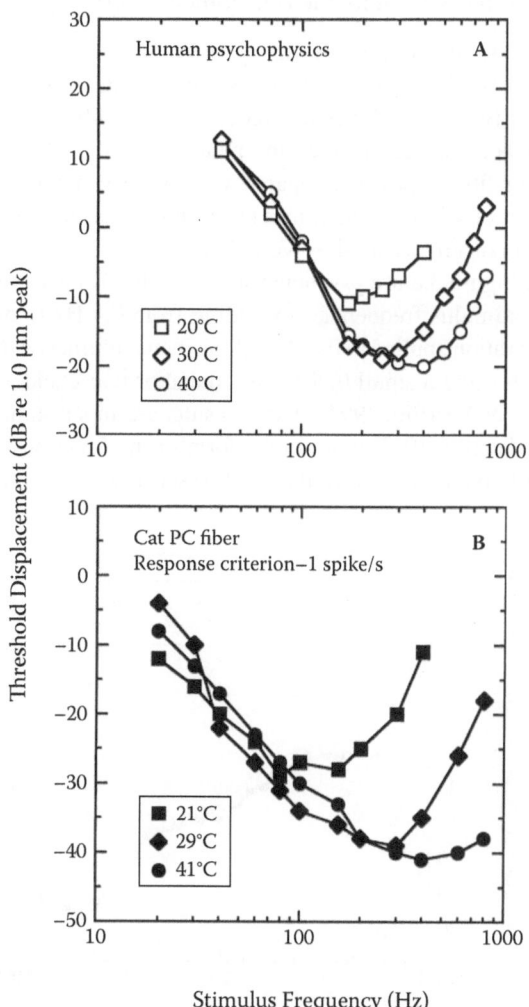

Figure 3.5 Threshold displacement as a function of stimulus frequency and temperature for psychophysical thresholds on the thenar eminence of a human (A) and for neural-activation thresholds of excised Pacinian corpuscles of a cat (B). From Bolanowski and Verrillo (1982).

ling evidence that the detection of high-frequency vibratory stimuli is mediated by PC nerve fibers. Furthermore, the shift in optimal sensitivity to higher stimulus frequencies as temperature increases has important implications for understanding the source(s) of the P-system frequency characteristics. The topic of frequency selectivity of channels is discussed later in this book.

Four Neural Systems Mediate the Detection of Vibratory Stimuli

In addition to PC and RA nerve fibers, there are SA I fibers innervating Merkel-neurite complexes and SA II fibers innervating SA II end organs. The role of the latter two types of mechanoreceptive nerve fibers with their associated receptors becomes clear when the neurophysiological tuning curve for each of the four fiber types is compared with the psychophysical thresholds for the detection of vibration measured over a much greater range of stimulus frequencies than originally used by Verrillo and others.

Plotted in Figure 3.6 are psychophysical thresholds measured over a very wide range of stimulus frequencies extending from 0.4 Hz to 500 Hz for the detection of vibration applied to the skin of the thenar eminence through a large 2.9-cm² contactor and a small 0.008-cm² contactor (Gescheider, Bolanowski, Hall, Hoffman, & Verrillo, 1994). These results are in substantial agreement with those of Verrillo (1963), both in the form of the functions obtained with large and small contactors and in the level of sensitivity demonstrated by the

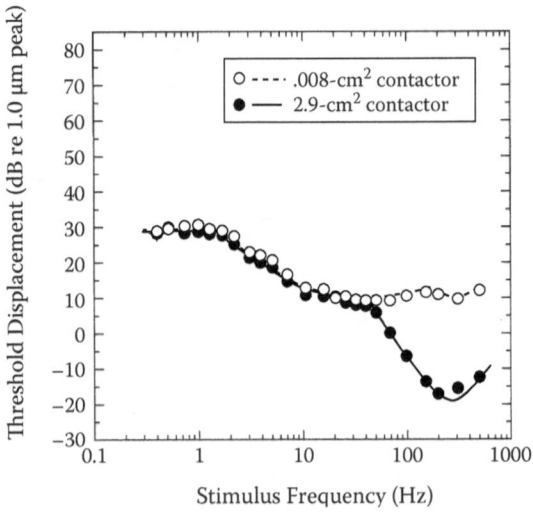

Figure 3.6 Detection thresholds as a function of stimulus frequency for stimuli applied to the thenar eminence through 0.008-cm² and 2.9-cm² contactors. From Gescheider, Bolanowski, Hall, Hoffman, & Verrillo (1994).

observers in both studies. The results of this study and those of Bolanowski et al. (1988), however, in extending the frequency range to much lower frequencies than used by Verrillo, demonstrate that the threshold-frequency function is more complex than originally obtained by Verrillo. In Verrillo's results, thresholds at frequencies between 25 Hz and 500 Hz were found to be high and nearly independent of stimulus frequency when stimuli were delivered through a small contactor. However, it is now known that at frequencies lower than 25 Hz the detection threshold first rises as the frequency of the stimulus becomes lower, and then between 0.4 Hz and 1.5 Hz the threshold becomes essentially independent of stimulus frequency. Thresholds measured with the large contactor are nearly identical to those measured with the small contactor when stimulus frequency is below 40 Hz. Thus, between 0.4 Hz and 40 Hz there is little or no spatial summation in the neural systems that mediate the detection threshold. In contrast, and in keeping with Verrillo's earlier results, thresholds above 40 Hz are lower when the stimulus is applied with the large contactor than when applied with the small one, thus showing that spatial summation occurs above 40 Hz.

How well can these expanded psychophysical measurements be accounted for by the frequency characteristics of the four fiber types known to mediate tactile sensation? To answer this question the tuning curves for each of the four mechanoreceptive fibers have been plotted in Figure 3.7 for comparison with the psychophysical thresholds. The tuning curves for the human SA I, SA II, and RA fibers were derived by Bolanowski et al. (1988) from the neurophysiological recordings of single nerve fibers innervating mechanoreceptors in the glabrous skin of the human hand (Johansson, Landström, & Lundström, 1982a). Tuning curves from monkey and cat fibers are also included. Each tuning curve represents the average intensity of vibration that causes the fiber to respond at a specified activity level (e.g., 1 spike per second, 4 spikes per second, etc.). To facilitate comparison of the psychophysical and neurophysiological frequency selectivity, the vertical position of each neural tuning curve has been adjusted up or down to achieve the best fit to the psychophysical threshold functions. The shapes of the tuning curves of the four fiber types closely match the shapes of the psychophysical threshold functions. Thus, we hypothesize that when the large contactor is used, thresholds above 40 Hz are determined by PC fibers (Figure 3.7A); but when the small contactor is used, thresholds above 100 Hz are determined by SA II fibers (Figure 3.7B). We further hypothesize that SA I fibers determine the psychophysical thresholds measured with either the small or the large contactor at low frequencies between 0.4 Hz and 1.5 Hz (Figure 3.7C), whereas RA fibers determine the thresholds between 1.5 Hz and 40 Hz when the large contactor is used and between 1.5 Hz and 100 Hz when the small contactor is used (Figure 3.7D).

The close correspondence between segments of the frequency characteristics of single nerve fibers innervating specific mechanoreceptors and segments

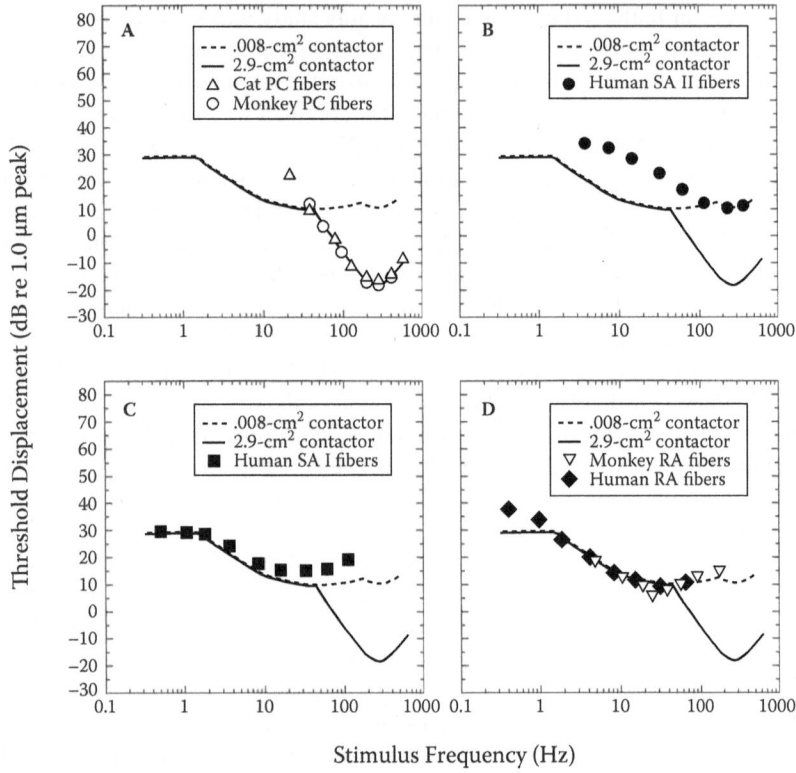

Figure 3.7 Detection-threshold functions of Figure 3.6 for 0.008-cm² and 2.9-cm² contactors compared with neural-activation thresholds. The human nerve-fiber data are taken from Johansson, Landström, & Lundström (1982a), the cat nerve-fiber data are from Bolanowski (1981), and the monkey nerve-fiber data are from Talbot et al. (1968).

of the psychophysical threshold function suggests that four neural systems (PC, RA, SA I, and SA II) mediate the detection of vibratory stimuli applied to glabrous skin. Furthermore, the frequency selectivity of each neural system appears to be determined solely by the frequency selectivity of the specific mechanoreceptor and associated nerve fiber that comprise the input stage to the system. It is important to point out, however, that the form of a neurophysiological tuning curve, which describes the intensity of a stimulus needed to produce a particular criterion neural response as a function of stimulus frequency, is highly dependent upon the particular neural-response criterion chosen by the investigator. For example, if action potentials are recorded and the response criterion is expressed in terms of spikes per second, then the shape of the tuning curve will change as the specified criterion firing rate is changed

(Bolanowski et al., 1988). Likewise, if one specifies the criterion as a certain number of spikes occurring during a burst of stimulation, then the shape of the tuning curve is determined by the number of spikes in the burst used as the response criterion (Bolanowski et al., 1988). Thus, in constructing the neurophysiological tuning curve to compare psychophysical and neurophysiological tuning, one must have a well-developed rationale for the choice of a neural-response criterion (Bolanowski et al., 1988). Ideally, one must attempt to construct tuning curves in which the neural-response criterion chosen matches the minimal neural response required by the observer to detect the stimulus (Bolanowski et al., 1988).

For example, the choice of one spike per stimulus burst as the neural-response criterion for RA fibers (Bolanowski et al., 1988) was based on evidence that observers are capable of detecting a single action potential on an RA fiber (Vallbo & Johansson, 1976). Specifically, when a single action potential is produced by electrical stimulation of an RA fiber, human observers usually report a sensation in the area of skin innervated by the fiber. Thus, the minimal neural response required for the detection of stimuli by the RA system appears to be a single spike on a single RA fiber. However, based on a population model of RA fibers, Güçlü and Bolanowski (2004) have determined that 5 to 10 RA fibers must be activated for stimulus detection to occur in the RA system. The choice of four spikes per stimulus burst as the neural-response criterion for PC fibers (Bolanowski et al., 1988) was based on the fact that to account for temporal summation in the PC system, the minimal neural response required for stimulus detection must be more than a single spike per stimulus burst but less than one spike for each sinusoidal indentation of the skin occurring within the stimulus burst (Van Doren, 1985). Although the exact minimal neural-response requirement for the PC system cannot be specified at this time, it is clear from examination of the tuning curves constructed from different neural-response criteria that the peripheral input to the PC system must be the activity of PC fibers. Furthermore, the tuning curves of the other three known fiber types cannot account for the magnitude of the high-frequency tuning of the PC system measured psychophysically.

A neural-response criterion of five spikes per second was chosen for SA II fibers (Bolanowski et al., 1988) because the minimal spike rate required for the observer to detect the stimulus must be sufficiently high to overcome the spontaneous activity that is inherent in these fibers. It appears that to be detected, neural activity in several SA II fibers must occur, inasmuch as spikes in a single fiber elicit no reports of conscious awareness from human observers (Ochoa & Torebjörk, 1983). Finally, the choice of a neural-response criterion of an action-potential firing rate of 0.8 spikes per second or greater for SA I fibers (Bolanowski et al., 1988) was based on the fact that the neural tuning curve for SA I nerve fibers matches the psychophysical-threshold frequency characteristic only when the criterion firing rate is greater than 0.8 spikes per second.

Frequency Selectivity of a Neural System Is Determined by Its Receptors

Because it is possible to identify the specific peripheral mechanoreceptive nerve fibers that constitute the input stage to each of the four neural systems, henceforth each system will be identified by its peripheral nerve fiber (PC, RA, SA I, or SA II). The frequency selectivity of each tactile neural system appears to be determined solely by the frequency selectivity of its receptors and their associated peripheral nerve fibers. Indeed, the close match between changes in the psychophysical and neurophysiological thresholds as the frequency of the stimulus is changed is quite remarkable. This constitutes strong evidence that frequency selectivity at the final stage of each of the four systems (namely, the psychophysical threshold function) is determined entirely by the frequency selectivity at its earliest stage (namely, receptors and associated nerve fibers), with no additional frequency selectivity occurring later in the central nervous system.

The principle that receptor characteristics alone can explain how psychophysical thresholds change as qualitative dimensions of the stimulus change is also seen in the visual system. In scotopic vision, changes in the detection threshold measured at the retina as the wavelength of light applied to rods is varied are determined exclusively by the spectral sensitivity of the rod receptor (Cornsweet, 1970). The same principle appears to apply in photopic vision as well, in which changes in psychophysical thresholds for detecting light measured at the fovea can be accounted for by the spectral sensitivities of each of three types of cones found exclusively in this part of the human retina (Cornsweet, 1970). Our analysis of the tactile system is consistent with the general principle that the filtering characteristic of a sensory system is determined primarily, if not exclusively, by the filtering characteristics of its receptors. Thus, in the sense of touch it is appropriate to identify each of the four neural systems by receptor type (see Figure. 3.1).

4

From Neural Systems to Information-Processing Channels

Introduction

It is clearly apparent that multiple neural systems—each with its own specific type of receptor and nerve fiber—mediate the detection of tactile stimuli of varying vibratory frequency. Furthermore, it appears that the minimal neural requirement for detecting the stimulus is unique for each neural system in ways that reflect other characteristics of the system, such as the capacity for spatial and temporal summation in the PC system and the presence of spontaneous activity in the SA II system. However, the fundamental question remains: do these neural systems constitute the inputs to separate information-processing channels, which serve as building blocks in the construction of complex tactile perceptions? To answer this question, it is necessary to demonstrate experimentally that: (1) at early stages each system independently extracts information to which it is optimally tuned, and in so doing is uninfluenced by changes in the neural activity in other systems resulting from factors such as adaptation, masking, and sensation-magnitude enhancement by prior stimulation; and (2) by combining their separate information, the channels interact at later stages to construct complex perceptual representations of tactile stimuli.

Sensation-Magnitude Enhancement Occurs Within but Not Across Channels

The discovery of enhancement of the perceived magnitude of a stimulus by the presentation of a prior stimulus within but not across tactile systems constituted the original evidence for the tactile-channel hypothesis first proposed by Verrillo and Gescheider (1975). Sensation-magnitude enhancement was demonstrated first in hearing (e.g., Zwislocki, Ketkar, Cannon, & Nodar, 1974; Zwislocki & Sokolich, 1974) and then in touch (e.g., Verrillo & Gescheider, 1975). In both the auditory and tactile psychophysical experiments, observers were required to adjust the intensity of a comparison stimulus so that its sensation magnitude matched that of a test stimulus. The comparison stimulus was presented 500 ms after the termination of the test stimulus. This was done both with and without a conditioning stimulus presented prior to the test stimulus, as illustrated in Figure 4.1A. The observer was instructed to attend only to the test stimulus when making the match. Enhancement was shown by the observer's adjusting the comparison stimulus to an intensity level higher when the test stimulus was preceded than when not preceded by the conditioning stimulus. The amount of enhancement was the difference in intensity of the matches made to the test stimulus with and without the conditioning stimulus.

As seen in Figure 4.1B, enhancement was found to be greatest when the conditioning and test stimuli were presented nearly simultaneously. Similar results have been reported for hearing (Zwislocki et al., 1974; Zwislocki & Sokolich, 1974). Thus, in both touch and hearing, the persisting neural effects of a brief stimulus may outlast the stimulus by as long as 400–500 ms and have an enhancing effect on the perceived magnitude of a second stimulus. But what is quite clear is that in both modalities this happens only when the frequencies of the conditioning and test stimuli are identical or similar. For example, Figure 4.1B shows that in tactile experiments, large amounts of enhancement are observed when the frequencies of the test and conditioning stimuli are the same—either 300 Hz or 25 Hz—but not when they are very different—25 Hz and 300 Hz. Thus, whereas enhancement occurs when the conditioning and test stimuli are presented within the same neural system (PC system at 300 Hz or RA system at 25 Hz), enhancement disappears when the first and second stimuli activate the RA and PC systems, respectively. The two systems do not appear to interact in a way that produces an enhancement effect.

The tactile results are analogous to the results of auditory experiments in which loudness enhancement is found when the frequencies of the two stimuli are in the same critical band but not when they are presented in different critical bands (Zwislocki et al., 1974; Zwislocki & Sokolich, 1974). As with critical bands in hearing, the PC and RA systems in touch appear to operate as information-processing channels capable of integrating the neural responses of two stimuli, provided the stimuli are presented within the same system. Indeed, the amount of enhancement can be precisely predicted by the degree to which two stimuli that differ in frequency stimulate the same channel (Gescheider, Verrillo, Capraro, & Hamer, 1977).

The occurrence of enhancement within but not across tactile systems raises the important question of where the phenomenon occurs within the nervous system. Because there is essentially no persisting neural activity in peripheral tactile nerve fibers following the termination of a vibratory stimulus

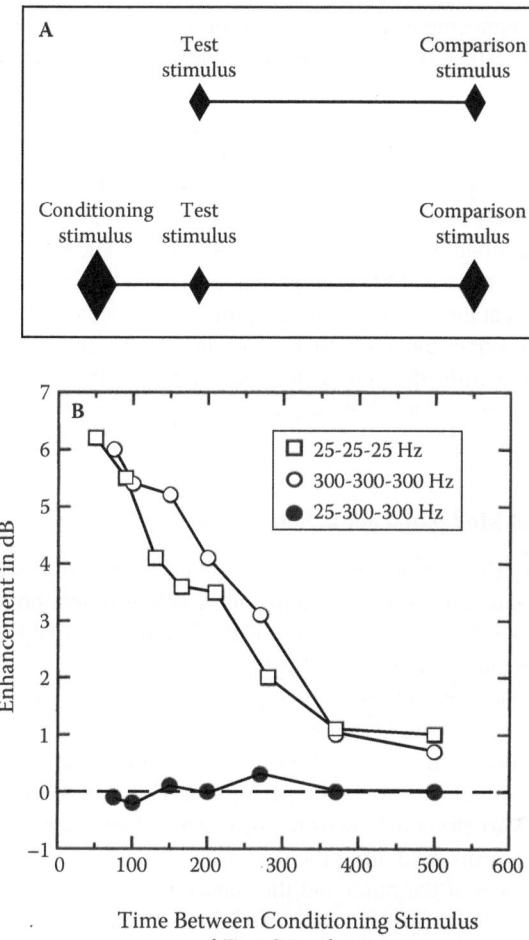

Figure 4.1 Paradigm for measuring enhancement in which the observer adjusts a comparison stimulus so that its sensation magnitude is equal to that of a test stimulus presented 500 ms earlier. Matches are made when the test stimulus is presented alone and when it is presented at a variable time after the presentation of a conditioning stimulus (A). Enhancement is expressed as the difference in dB between matches to the test stimulus when presented after a conditioning stimulus and matches to the test stimulus when presented alone (B) (Verrillo & Gescheider, 1975).

(Bolanowski, 1981), we conclude that the phenomenon of enhancement cannot be accounted for by a process of summation in the peripheral nerve fibers in which the persisting neural activity produced by a conditioning stimulus adds to the neural activity produced by a test stimulus presented later. Hence, the process of enhancement must occur within the central nervous system. This conclusion is further supported by the occurrence of enhancement across the body (Gescheider & Verrillo, 1982). When the conditioning stimulus is presented on one hand and the test stimulus is presented on the homologous site of the contralateral hand, substantial enhancement occurs. Under these conditions there is no opportunity for the neural effects of the conditioning and test stimuli to interact in the peripheral nervous system, and therefore these effects must occur at a higher level in the nervous system. The finding that enhancement occurs within a tactile system but not between tactile systems, along with the evidence that enhancement must occur within the central nervous system, constitutes powerful evidence that tactile neural systems behave as independent channels within the central nervous system, while having their origins in specific receptors and their associated nerve fibers within the peripheral nervous system.

Multichannel Model of Tactile Sensitivity

The discovery that tactile neural systems behave like information-processing channels with regard to the phenomenon of enhancement provides the major theoretical basis for a multichannel model of tactile sensitivity. The fundamental idea of the multichannel model is that different aspects of the tactile stimulus are independently processed by separate information-processing channels, each with its own specific type of sensory receptor and associated nerve fiber, and it is the combined activity of the channels within the central nervous system that determines overall tactile perception. The genesis of this concept was the work of Verrillo (1963) with his discovery of two separate and independent neural systems mediating the detection of mechanical stimuli applied to the glabrous skin of the hand and the subsequent discovery that these neural systems exhibit the properties of information-processing channels (Verrillo & Gescheider, 1975). The multichannel model further evolved with the discovery that the original NP system is actually comprised of three distinct NP systems (Bolanowski et al., 1988; Capraro et al., 1979; Gescheider, Sklar, Van Doren, & Verrillo, 1985), each possessing a specific type of nerve fiber and receptor type (Bolanowski et al., 1988). These are the NP I system with its RA fibers and Meissner corpuscles, the NP II system with its SA II fibers and SA II end organs, and the NP III system with its SA I fibers and Merkel-neurite-complex receptors. Henceforth these systems will be referred to by their nerve-fiber type and not by their original NP designation.

The multichannel model as it is currently formulated to explain the detection of vibratory stimuli of varied frequency delivered to the thenar eminence through a large 2.9-cm^2 contactor and through a small 0.008-cm^2 contactor is illustrated in Figure 4.2. By comparing the model for the 2.9-cm^2 and 0.008-cm^2 contactors, it can be seen that the frequency-selectivity functions of the

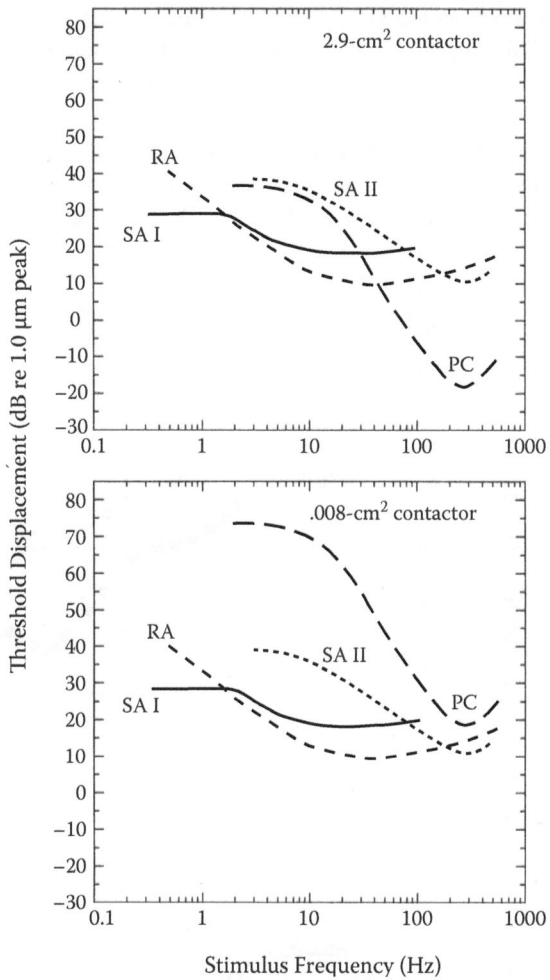

Figure 4.2 Multichannel model of tactile sensitivity. The activation thresholds of the SA I, SA II, RA, and PC channels are plotted as a function of stimulus frequency for stimuli applied to the glabrous skin of the hand through a 2.9-cm² or a 0.008-cm² contactor. The frequency-selectivity functions for the channels are taken from Gescheider, Bolanowski, and Hardick (2001).

RA, SA I, and SA II channels are unaffected by changing the size of the skin area stimulated. This is because these three channels are incapable of spatial summation. Consequently, the thresholds of these channels at any particular stimulus frequency remain the same as the size of the contactor varies. In sharp contrast, the PC channel's threshold at any stimulus frequency becomes

dramatically lower as the size of the contactor is increased from 0.008 cm²
to 2.9 cm². This improvement in sensitivity as the stimulated area of skin is
increased, referred to as spatial summation, is found only in the PC channel.

Figure 4.3 shows that the model accounts remarkably well for psychophysical thresholds obtained with large and small contactors at all stimulus frequencies. With both the large and small contactors, thresholds between 0.4 Hz
and 2 Hz are determined by the SA I channel. With the large contactor the RA

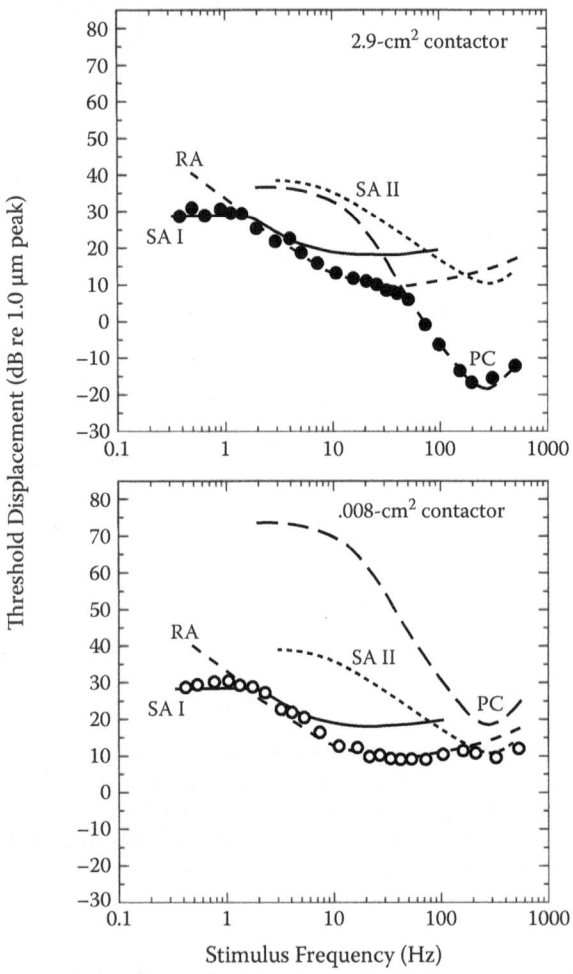

Figure 4.3 Multichannel model of tactile sensitivity as seen in Figure 4.2 but with
detection-threshold data added for stimuli applied to the thenar eminence through
2.9-cm² and 0.008-cm² contactors. Detection-threshhold data from Gescheider,
Bolanowski, Hall, Hoffman, and Verrillo (1994).

channel determines thresholds between 2 Hz and 40 Hz and the PC channel determines thresholds above 40 Hz. With the small contactor the thresholds of the PC channel become so high that psychophysical thresholds are not determined by this channel at any frequency. Instead, the RA channel determines thresholds between 2 Hz and 100 Hz and the SA II channel determines thresholds above 100 Hz.

The frequency selectivity of each channel was determined psychophysically by selective adaptation or masking of the other channels. With either of these procedures, the frequency-selectivity function of a channel can be examined over a very wide range of stimulus frequencies. This is accomplished by elevating the thresholds of all other channels through the use of appropriate adapting or masking stimuli. The success of this approach is predicated on two fundamental principles in sensory science: (1) the psychophysical threshold for detecting a stimulus is always determined by the neural system with the lowest threshold for activation by that stimulus; and (2) adaptation and masking can act selectively by elevating the thresholds of one neural system without affecting the others. Both of these principles clearly operate in the tactile sensory system (Bolanowski et al., 1988; Gescheider et al., 1979; Gescheider & Verrillo, 1979; Gescheider et al., 1982; Gescheider et al., 2001; Gescheider et al., 2002; Hollins et al., 1990; Labs et al., 1978; Verrillo & Gescheider, 1977) and permit precise experimental determination of the frequency selectivity of each of the four channels.

The Psychophysical Tuning Curve

Of all the procedures used to isolate and study tactile channels, the procedure used in audition to obtain psychoacoustic tuning curves (Vogten, 1974; Zwicker, 1974; Zwicker & Schorn, 1978; Florentine, Buus, Scharf, & Zwicker, 1980; Bacon & Jesteadt, 1987) provides the most effective means of determining the tuning of a tactile channel over a wide range of stimulus frequencies (Gescheider et al., 2001; Labs et al., 1978). In this method, a test stimulus of fixed frequency that selectively activates a single channel is presented at a clearly detectable level well above the detection threshold. Masking stimuli of varied frequency are then presented, and the intensity of each masking stimulus needed to render the test stimulus barely detectable is recorded. The psychophysical tuning curve is a graphic representation of these masking-stimulus intensities plotted as a function of their frequency and provides a measure of the frequency selectivity of the channel. In a variation of this procedure, the tuning curve for a channel is obtained by determining the intensities of adapting rather than masking stimuli of various frequencies needed to maintain a test stimulus barely detectable within the single channel under investigation (Gescheider et al., 2001).

This method for measuring the frequency selectivity of a channel completely isolates the channel over the entire frequency range within which it is responsive to stimuli. The channel of interest is isolated by selecting a stimulus of a particular frequency and intensity and presenting it through a contactor of appropriate size so that it activates only that channel. Throughout a testing session, the frequency and intensity of the test stimulus remain unchanged, but the intensity of the masking or adapting stimulus is adjusted to maintain the test stimulus at a constant level of detectability (e.g., detecting the test stimulus correctly 75 percent of the time in a two-alternative forced-choice task). The implicit assumption underlying this method is that neural activity in one channel does not mask or adapt neural activity in another. Therefore, although a masking or adapting stimulus may alter the sensitivity of several channels at the same time, the psychophysical tuning curve represents the frequency selectivity of the single channel isolated through use of the appropriate test stimulus.

In Figure 4.4, the intensity of the adapting stimulus required to maintain the test stimulus detectable 75 percent of the time is plotted as a function of the frequency of the adapting stimulus for test stimuli that exclusively stimulate the PC channel (250 Hz, 1.5-cm² contactor), the RA channel (22 Hz, 1.5-cm² contactor), and the SA II channel (250 Hz, 0.008-cm² contactor). (See Gescheider

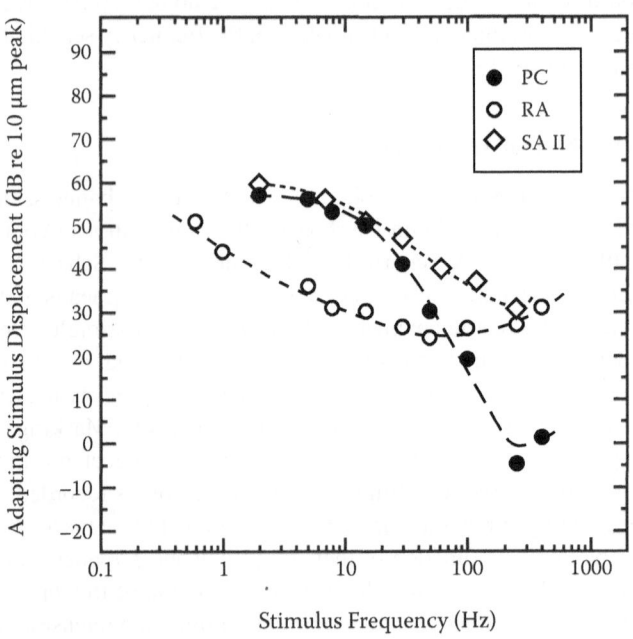

Figure 4.4 Psychophysical tuning curves for the PC, RA, and SA II channels measured by an adaptation procedure. From Gescheider et al. (2001).

et al., 2001.) These psychophysical tuning curves were used to construct the frequency-selectivity functions of the PC, RA, and SA II channels in the multichannel model presented in Figure 4.2. The frequency-selectivity function for the SA I channel was determined by using the masking procedure. The SA I channel was isolated by using a 0.7-Hz test stimulus, and the psychophysical tuning curve was obtained by determining the intensity of masking stimuli at various frequencies needed to maintain the 0.7-Hz test stimulus detectable 75 percent of the time (Bolanowski et al., 1988). The SA I psychophysical tuning curve (Figure 4.5) provides the basis for constructing the frequency-selectivity functions of the SA I channel seen in Figures 4.2 and 4.3. Because the psychophysical tuning-curve procedure used to measure the frequency selectivity of a channel is valid only if masking and adaptation occur within but not across channels, the validity of this assumption must be thoroughly examined.

Testing the Multichannel Model Through Experiments on Adaptation and Masking

Traditionally, the question of whether sensory neural systems such as the PC, RA, SA I, and SA II systems of somatosensation function as channels has been answered through experiments in which adaptation or masking procedures are employed. The objective of these experiments is to determine whether the sensitivity of a particular neural system is affected by compromising the sensitivity of another neural system through adaptation or masking. For example,

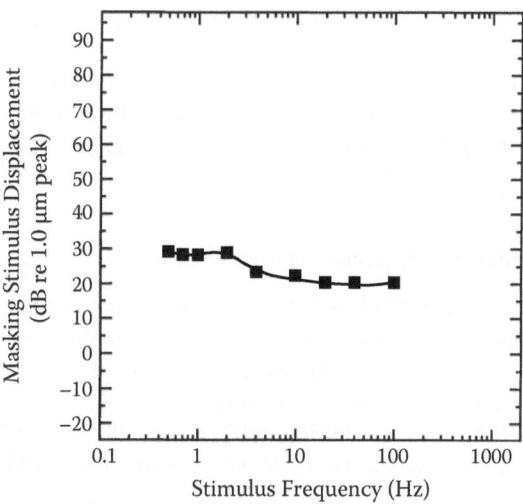

Figure 4.5 Psychophysical tuning curve for the SA I channel measured by a masking procedure. From Bolanowski et al. (1988).

in vision, spatial-frequency channels were discovered by demonstrating that adaptation to a particular spatial frequency through extended exposure of the eye to a grid pattern of a specific frequency, although elevating the contrast-sensitivity threshold at that frequency, has little effect on thresholds at higher or lower spatial frequencies (Campbell & Robson, 1968). These results indicate that the effects of adaptation to a specific spatial frequency are limited to frequencies that are the same or similar to that of the adapting stimulus. From this finding it was concluded that the visual system has a number of independent spatial-frequency channels, each tuned to a specific range of frequencies (Blakemore & Campbell, 1969; Campbell & Robson, 1968). Thus, spatial-frequency channels consist of sets of detectors distributed across the visual field, each tuned to a narrow range of frequencies.

Channels referred to as critical bands have been identified in the auditory system by a number of techniques, including masking experiments. Critical bands are frequency ranges of the acoustic stimulus within which the detectability of a stimulus can be impaired (masked) by the presentation of another stimulus within the same band of frequencies (Feldtkeller & Zwicker, 1956; Fletcher, 1940; Greenwood, 1961). The channel hypothesis in the auditory system is further supported by the finding that masking does not occur when the masking and test stimuli are presented in different critical bands.

In the tactile sensory system, the fundamental question is whether adaptation and masking occur within but not across tactile neural systems. This question can be answered by conducting experiments in which the threshold for detecting a test stimulus is measured either following exposure of the skin to adapting stimuli of varied intensity or during the presentation of masking stimuli of varied intensity. If tactile neural systems behave as channels, then the results of such experiments should reveal that the threshold for detecting a test stimulus is elevated by an adapting or masking stimulus only when the adapting or masking stimulus activates the same neural system as the one detecting the test stimulus.

Adaptation Reveals the Existence of Tactile Channels

In somatosensory research it has long been known that exposing the skin to prolonged stimulation can cause a loss of sensitivity in the area within which an adapting stimulus is presented (Berglund & Berglund, 1970; Cohen & Lindley, 1938; Gescheider et al., 1978; Gescheider et al., 1979; Gescheider & Wright, 1968, 1969; Goble & Hollins, 1993; Hahn, 1966, 1968; Hollins et al., 1990; Verrillo & Gescheider, 1977; Wedell & Cummings, 1938). Such sensory adaptation results from a reduction in the responsivity of both first-order mechanoreceptive afferents (Lundström & Johansson, 1986; Leung, Hsiao, & Johnson, 1994) and central nervous system neurons (O'Mara, Rowe, & Tarvin, 1988). It is also now known that in the case of vibratory stimuli the amount of adaptation depends greatly on whether the frequencies of the adapting and test stimuli are similar or different (Gescheider et al., 1978; Gescheider et al., 1979; Hahn, 1968; Hollins et al., 1990; Verrillo & Gescheider, 1977).

Shown in Figure 4.6A are the results of an experiment in which an adapting stimulus with a frequency of 250 Hz, chosen to adapt optimally the PC channel, was applied for 10 minutes to the thenar eminence, following which the threshold for detecting either a 140-Hz or 20-Hz test stimulus was measured (Gescheider et al., 1979). When the test stimulus was 140 Hz, well within the frequency range of the PC channel, detection thresholds began to rise as soon as the intensity of the 250-Hz adapting stimulus exceeded the 250-Hz detection threshold. In sharp contrast, the threshold data obtained when the adapting stimulus was 250 Hz and the test stimulus was 20 Hz were in close agreement with the function predicted from the independent-channel hypothesis (solid line). Specifically, thresholds for the detection of a 20-Hz test stimulus measured after 250-Hz adaptation were not significantly elevated from those measured prior to adaptation until the intensity of the 250-Hz adapting stimulus was more than 25 dB above the 250-Hz detection threshold. This intensity value of the 250-Hz adapting stimulus at which the threshold of the 20-Hz test stimulus begins to rise can be predicted from the frequency-selectivity functions of the RA and PC channels fitted to the unadapted detection thresholds as shown in Figure 4.6B. It is clear that at 250 Hz the threshold of the RA channel is 25 dB above that of the PC channel—a value that coincides exactly with the intensity of the 250-Hz adapting stimulus minimally needed to elevate the threshold for detecting the 20-Hz test stimulus by the RA channel (Figure 4.6A). From these results it follows that adaptation occurs within but not across tactile channels. There is no evidence that adaptation of the PC channel compromises the sensitivity of the RA channel measured at 20 Hz. Instead, only when the 250-Hz adapting stimulus was sufficiently intense to exceed the threshold of the RA channel at 250 Hz was the 20-Hz threshold elevated.

Further evidence that the 140-Hz and 20-Hz stimuli were detected by different channels is the fact that the average rate at which thresholds are elevated as the intensity of the adapting stimulus is increased is substantially different for the two frequencies. In the PC channel, thresholds increased at a rate of 4 dB for every 10-dB increase in the intensity of the adapting stimulus, whereas thresholds in the RA channel increased at the higher rate of 6 dB per 10-dB increase in the intensity of the adapting stimulus. Thus, in addition to its capacity for temporal and spatial summation and its more highly tuned frequency characteristic, the PC channel differs from the RA channel by being less affected by sensory adaptation.

The results of a complementary experiment in which the adapting stimulus was 10 Hz, well within the frequency range of the RA channel, are consistent with those described above, again showing that adaptation occurs within but not across channels (Verrillo & Gescheider, 1977). As seen in Figure 4.7A,

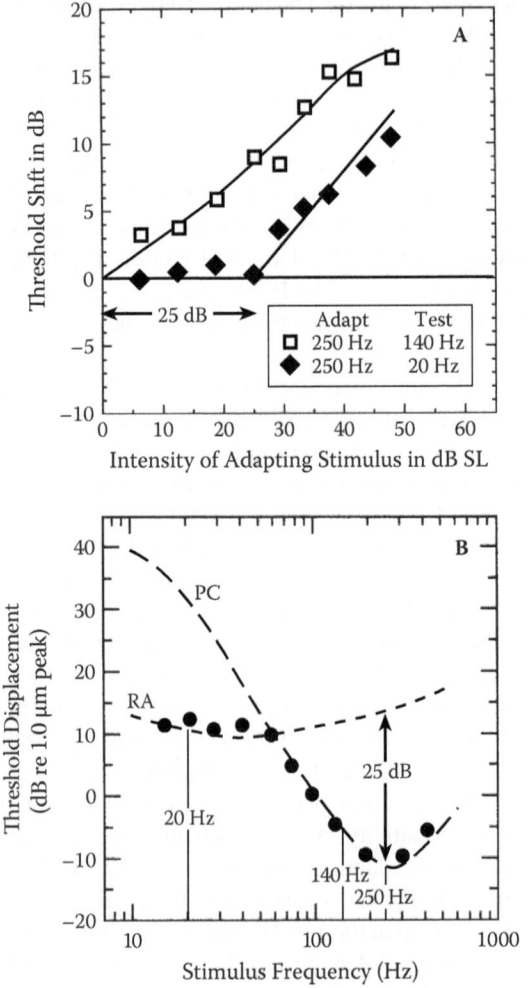

Figure 4.6   Threshold shift in dB relative to the unadapted threshold resulting from exposure to a 250-Hz adapting stimulus plotted as a function of the intensity of the adapting stimulus (A). Threshold displacement as a function of stimulus frequency compared with the frequency-selectivity functions of the PC and RA channels (B). From Gescheider et al. (1979).

thresholds for detecting the 15-Hz test stimulus began to rise as soon as the intensity of the adapting stimulus exceeded the 10-Hz detection threshold. The rate at which thresholds for detecting the 15-Hz test stimulus increased as the intensity of the adapting stimulus increased was approximately 6 dB per

10-dB increase in adapting stimulus intensity—a value corresponding exactly to that seen when the adapting stimulus was 250 Hz and the test stimulus was 20 Hz (Figure 4.6A). Thus, under very different stimulus conditions, but ones where the detection threshold is mediated by the RA channel, the rate of adaptation is the same.

Figure 4.7B shows that the 10-Hz adapting stimulus must be at least 26 dB above the detection threshold at 10 Hz to exceed the threshold of the PC

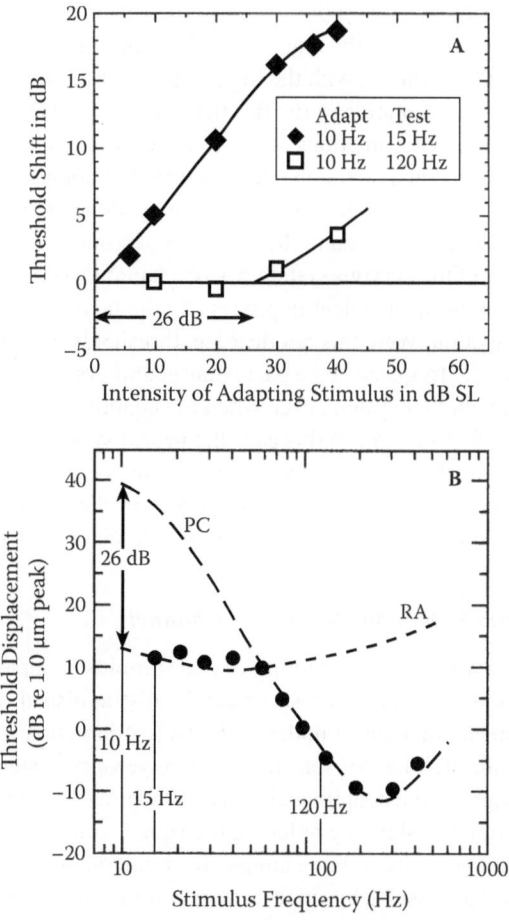

Figure 4.7 Threshold shift in dB relative to the unadapted threshold resulting from exposure to a 10-Hz adapting stimulus plotted as a function of the intensity of the adapting stimulus (A). Threshold displacement as a function of stimulus frequency compared with the frequency-selectivity functions of the PC and RA channels (B). From Verrillo and Gescheider (1977).

channel, and as seen in Figure 4.7A, it is at this intensity level that the adapting stimulus begins to elevate the threshold of the PC-mediated threshold for detecting a 120-Hz stimulus. Furthermore, the slope of the adaptation function above 26 dB is approximately 4 dB per 10-dB increase in the intensity of the adapting stimulus. Thus, the slopes of the PC-channel adaptation functions are essentially identical in the two experiments in which the frequencies of the adapting stimuli were very different. From the results of these selective-adaptation experiments, it can be concluded that: (1) adaptation occurs within but not across tactile channels; (2) the slopes of the functions describing changes in threshold as a function of the intensity of the adapting stimulus are different for the RA and PC channels, with the higher slope for the RA channel indicating a higher rate of adaptation in the RA than in the PC channel; and (3) from the frequency-selectivity functions of the RA and PC channels, in conjunction with the hypothesis that adaptation occurs within but not across channels, it is possible to predict precisely the minimal intensity value of an adapting stimulus that will produce adaptation in the detection of stimuli of varied frequency. Thus, at the level of the nervous system where adaptation occurs, it appears that the channels remain independent in processing the tactile information needed for stimulus detection. Were this not the case, thresholds for detecting a stimulus of some specific frequency by a particular neural system would be elevated by adaptation in another system, even when the adapting stimulus is below the threshold of the first system. In this case the neural systems would not be acting as independent channels, because adaptation in one system could affect the sensitivity of another. This clearly did not occur in these experiments, which provides strong support for the independent tactile-channel hypothesis.

Masking Occurs Within but Not Across Channels

Masking occurs when the detectability of one stimulus is impaired by the presentation of another stimulus. Masking can be distinguished from adaptation by the mechanisms thought to underlie the two. Adaptation raises the detection threshold through loss of sensitivity in the sensory system as a result of extended exposure to the adapting stimulus. In contrast, masking is thought to raise the detection threshold by reducing the signal-to-noise ratio by presentation of the masking stimulus. For example, in simultaneous masking the signal must be detected against background stimulation such as noise, and in forward masking it is the persisting effects of the masking stimulus against which the signal must be detected. The more intense the masking stimulus, the lower the signal-to-noise ratio describing the relative intensity levels of the signal and the masking stimulus, and the more difficult it becomes to detect the signal of a particular intensity. Whatever the precise mechanism(s) responsible for masking, it is a phenomenon that is pervasive in sensory systems, including the tactile sensory system, and fortunately it affords the opportunity to test further the tactile-channel hypothesis.

In one experiment on tactile masking (Gescheider et al., 1982), sinusoidal vibratory stimuli applied through a 2.9-cm² contactor to the thenar eminence were detected against a vibratory noise background. The 700-ms masking stimulus was narrow-band noise centered at 275 Hz, and the frequency of the 300-ms test stimulus temporally centered in the masking noise was either 300 Hz or 15 Hz. Masking functions, which show the shift in the threshold for detecting the test stimulus resulting from presentation of the masking stimulus, are plotted in Figure 4.8A. These functions provide strong evidence, as the following analysis shows, that multiple independent channels exist in the tactile system.

Consider the hypothesis that only a single channel responds to vibratory stimulation. If this were true, then all masking functions, regardless of test-stimulus frequency, would have the same form. For example, if only the PC channel responded to vibration, then narrow-band noise centered at 275 Hz should elevate the threshold for detecting 15-Hz and 300-Hz test stimuli by exactly the same amount because the masking stimulus, in effect, would uniformly elevate the entire frequency threshold curve of the PC channel by the same amount. However, if detection of the 15-Hz stimulus is mediated by a different channel, such as the RA channel, then the high-frequency noise-masking stimulus should elevate the detection threshold only when its intensity has become sufficiently high to excite this channel. This prediction is based on the assumptions that: (1) two or more channels with different frequency characteristics mediate the detection of vibrotactile stimuli; (2) at any particular frequency of the test stimulus, the psychophysical threshold is always determined by the channel with the lowest activation threshold; and (3) neural activity in one channel cannot mask neural activity in another. The first two assumptions are strongly supported by the psychophysical and physiological findings reviewed earlier in this monograph. Measuring the psychophysical threshold for detection of a 15-Hz test stimulus as a function of the intensity of a masking stimulus of much higher frequency is a strong test of the third assumption, namely, that there is no cross-channel masking. Specifically, if there are no cross-channel effects, masking of the 15-Hz stimulus by high-frequency noise should occur only when the intensity of masking noise exceeds the threshold of the RA channel. As seen in Figure 4.8B, the threshold of the RA channel used to detect a 15-Hz test stimulus is 29 dB higher than that of the PC channel when the masking stimulus is 275 Hz. Consequently, the masking function for the 15-Hz test stimulus should show no shift in threshold for intensities of the masking stimulus between 0 and 29 dB above the threshold for detection of the masking stimulus. Above 29-dB sensation level (SL) masking should occur because the intensity of the 275-Hz masking stimulus is now sufficiently high to activate the RA channel used to detect the 15-Hz test stimulus, and the

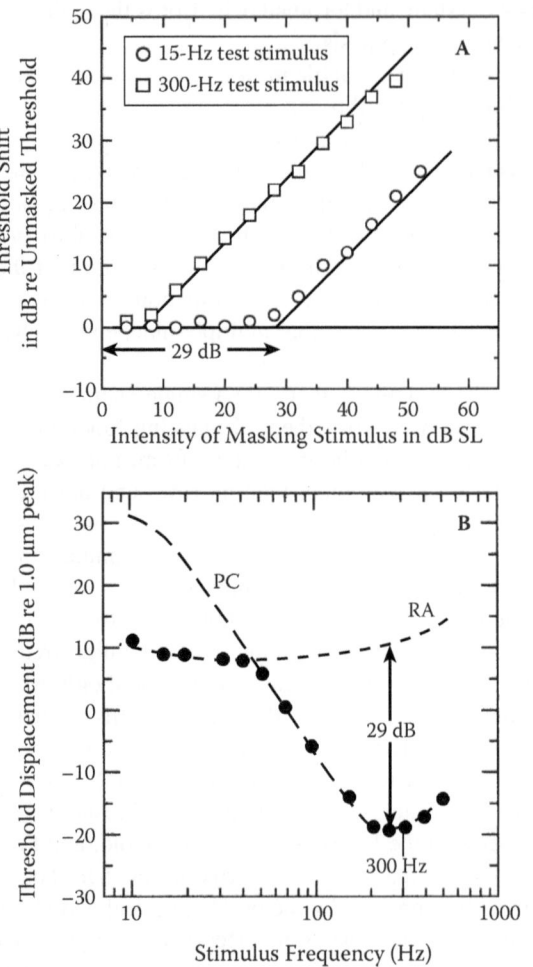

Figure 4.8 Masking expressed as a shift in the threshold for detecting a 300-Hz or a 15-Hz test stimulus attributable to a masking stimulus plotted as a function of the intensity of the masking stimulus (A). Detection thresholds plotted as a function of stimulus frequency. The frequency-selectivity functions of the RA and PC channels are represented as dashed curves (B). From Gescheider et al. (1982).

slope of the function should abruptly change from 0 to 1. The results shown in Figure 4.8A fully confirm this prediction.

Now consider a more complex prediction for the masking function when the test stimulus is 80 Hz. First, consider the form of the masking function in Figure 4.8A for the PC channel obtained when the test stimulus was 300 Hz.

In this case masking does not begin to occur until the 275-Hz masking noise is 6 dB above the detection threshold for the noise. Above this point, masking increases with further increases in the intensity of the noise, and the function becomes linear with a slope of 1 (Gescheider et al., 1982). When the test stimulus is 80 Hz, it is predicted from the 300-Hz results that the detection of the test stimulus by the PC channel will not be affected by the masking stimulus until its intensity is 6 dB above the threshold for detecting the noise (Figure 4.9A and B). In these figures, N represents the short range of intensi-

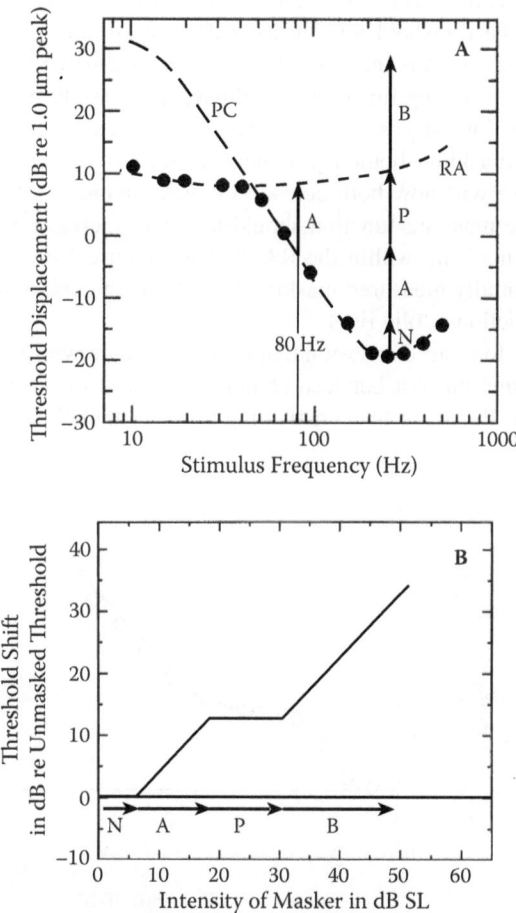

Figure 4.9 Detection thresholds as a function of stimulus frequency with frequency-selectivity functions of the RA and PC channels represented as dashed curves (A). Predicted masking function for an 80-Hz test stimulus. Masking is expressed as a shift in the detection threshold attributable to presentation of the masking stimulus (B). From Gescheider et al. (1982).

ties between 0 and 6 dB where the masking noise should not produce any shift in the threshold for detecting the test stimulus. Above N there is a range A of masking-noise intensities between 6-dB and 18-dB SL in which both the masking stimulus and the 80-Hz test stimulus fall below the threshold of the RA channel. Over this intensity range, the masking function should be linear with a slope of 1, representing masking within the PC channel. When the intensity of the masking noise rises above 18-dB SL, the threshold of the PC channel at 80 Hz will have been elevated above the threshold of the RA channel, and the RA channel rather than the PC channel will henceforth detect the test stimulus. If there is no cross-channel masking, there should be no further elevation of the threshold for detecting the test stimulus as the intensity of the masking stimulus is increased from 18-dB to 29-dB SL. Over this intensity range, the masking function should have a plateau P. When the masking-noise stimulus intensity is raised further (B) and exceeds the threshold of the RA channel, masking should again increase because the test stimulus and the masking noise will now both activate the RA channel. Between 29-dB and 50-dB SL the masking function should therefore increase with a slope of 1, representing masking within the RA channel. Figure 4.10 shows clearly that the experimentally measured masking function is in close agreement with all of these predictions (solid line).

It is therefore our conclusion that masking occurs within both the PC and the RA channel but not between channels. A masking stimulus that causes activity in the PC nerve fibers activating the PC channel can mask the detec-

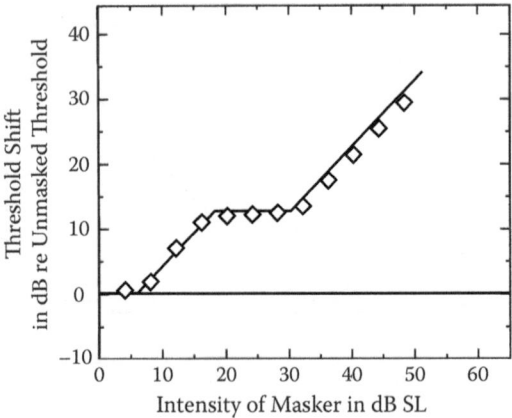

Figure 4.10 Masking function for an 80-Hz test stimulus in which masking expressed as a shift in the detection threshold attributable to the presentation of the masking stimulus is plotted as a function of the intensity of the masking stimulus. From Gescheider et al. (1982).

tion of test stimuli by Pacinian corpuscles. Likewise, neural activity in RA fibers can mask the detection of stimuli by Meissner corpuscles. Neural activity in one type of receptor, however, does not mask the detection of stimuli by the other. Instead, the Pacinian corpuscles and their associated PC nerve fibers and Meissner corpuscles with their associated RA fibers provide input to separate information-processing channels.

Testing the Multichannel Model Through Experiments on Sensory Learning

It has been conclusively shown that performance in a sensory-discrimination task improves with practice (Bolanowski, Hall, Makous, & Merzenich, 1995). An example of sensory learning in discriminating changes in stimulus intensity is as follows: Observers were required to indicate which of two sequentially presented time intervals contained the more intense vibrotactile stimulus applied to the thenar eminence. Feedback was given after each judgment. The intensity-difference limen (DL) was defined as the intensity difference that could be correctly discriminated on 75 percent of the presentations. In Figure 4.11 the average DL is plotted as a function of the number of practice sessions when the stimulus was isolated within the SA I and RA channels (20-Hz stimulus) and when it was isolated within the PC channel (250-Hz stimulus). In

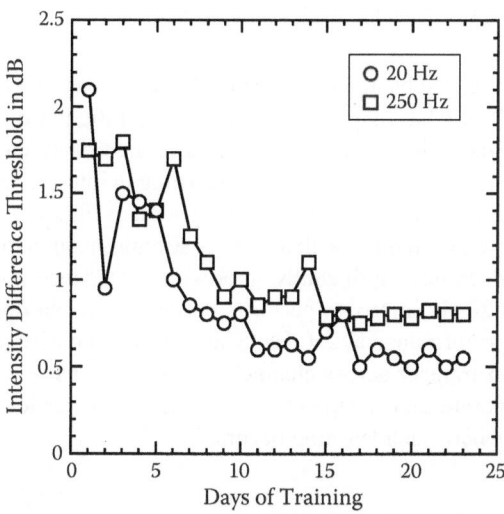

Figure 4.11 Effects of practice on intensity-difference thresholds for 20-Hz and 250-Hz stimuli expressed as the difference in dB between two just-discriminable stimuli presented to the thenar eminence. From Bolanowski et al. (1995).

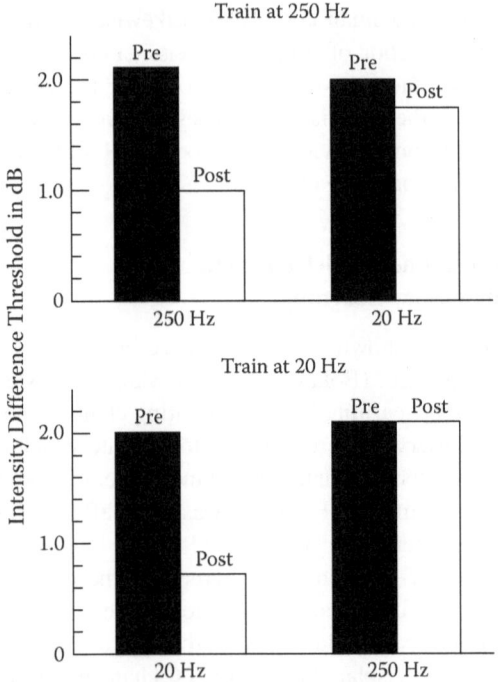

Figure 4.12   Pre- and post-training intensity-difference thresholds for the trained and untrained channels. From Bolanowski et al. (1995).

both cases learning is substantial, with the DLs at the end of training becoming less than half of what they were at the start, and the improvement in performance is gradual, with most of it occurring in the first 10 days of training. The experiment also revealed that sensory learning in intensity discrimination does not occur across tactile channels. Examination of Figure 4.12 reveals that training within the PC channel with a 250-Hz stimulus results in improvement at 250 Hz, but does not significantly improve performance at 20 Hz. Similarly, training at 20 Hz improves performance at 20 Hz, but does not significantly improve performance at 250 Hz. The fact that intensity-discrimination learning does not transfer across channels provides further evidence that the channels are separate and independent, even at the higher levels within the nervous system where such learning occurs.

5
Properties of Tactile Channels

Introduction

The evidence from psychophysical experiments on sensation-magnitude enhancement, adaptation, masking, and sensory learning strongly indicates that the four tactile neural systems behave as independent information-processing channels. Having established this, it now becomes important to determine if there are properties of the channels that distinguish one from another.

The Frequency Selectivity of Channels

How well can the frequency selectivity of channels be accounted for by the frequency selectivity of the four types of tactile receptors and their nerve fibers? To answer this question, the neurophysiological tuning curves plotted in Figure 3.7 are again plotted in Figure 5.1 for comparison with the psychophysical frequency-selectivity functions of each of the four channels. To facilitate these comparisons, the vertical position of each neurophysiological tuning curve has been set to best fit the positions of the psychophysical frequency-selectivity functions. As was true when examining the narrower ranges of the frequency-selectivity functions of the individual neural systems, there is close correspondence between the neurophysiological tuning curves and the frequency-selectivity functions measured psychophysically. Now, however, the correspondence between neurophysiological and psychophysical frequency selectivity is shown over a greatly expanded range of frequencies. These comparisons further strengthen the argument that it is the frequency selectivity of receptors, with their associated nerve fibers, that exclusively determines the frequency sensitivity of the neural system (channel) as a whole. This means that there are no higher centers in the central nervous system that have any effect on the basic frequency selectivity of the channels.

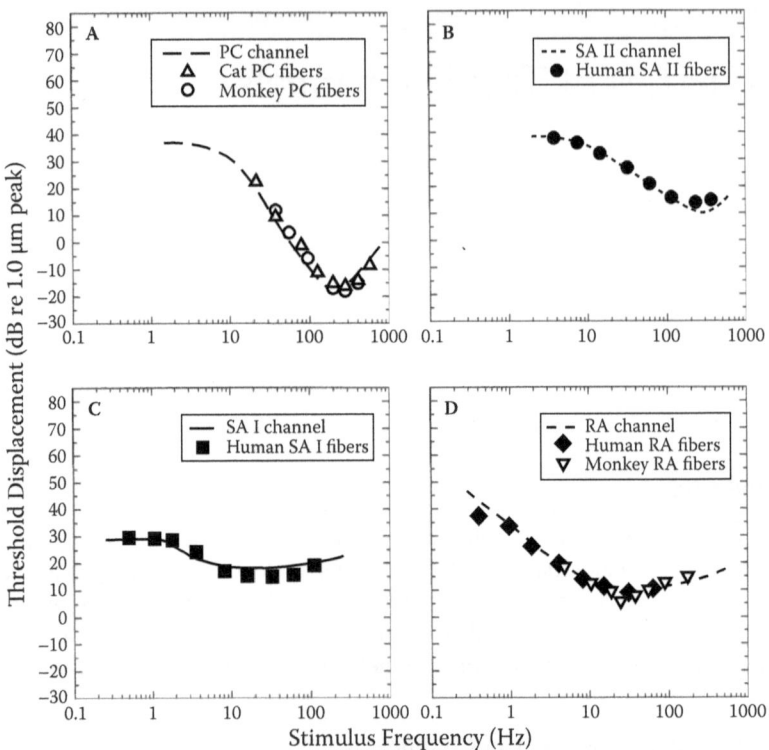

Figure 5.1 Psychophysical frequency-selectivity functions and neural-activation thresholds for each of the four channels, plotted as a function of stimulus frequency. Neural data points are the same as shown in Figure 3.7.

As discussed earlier, in constructing a neurophysiological tuning curve, the neural-response criterion is the same for all stimulus frequencies, namely, one spike per stimulus burst for RA fibers, four spikes per stimulus burst for PC fibers, a spike rate greater than 0.8 spikes per second for SA I fibers, and a spike rate of five spikes per second for SA II fibers. The fact that these tuning curves were constructed with the neural-response criterion held constant over the entire stimulus-frequency range, in conjunction with the fact that their forms essentially match the forms of the psychophysical frequency-selectivity functions of the channels, demonstrates a fundamental principle of the tactile sensory system. When the detectability of a stimulus determined psychophysically is held constant at all stimulus frequencies (in this case, the observer being correct 75 percent of the time in a two-alternative forced-choice task), the response of the peripheral nerve fibers mediating the observer's behavior is identical at all stimulus frequencies. For example, as seen in Figure 5.1D, a

1-Hz stimulus must be approximately 20 dB higher in intensity (10 times the amplitude and 100 times the energy) than a 30-Hz stimulus to be detected. However, the close correspondence between the forms of the frequency-selectivity curves measured psychophysically and neurophysiologically is strong evidence that the same neural response is required for stimulus detection at both frequencies. This finding, taken with the finding that human observers can detect a single spike induced in RA fibers (Vallbo & Johansson, 1976), suggests that a single spike in one or more RA fibers results in stimulus detection at all stimulus frequencies, although the stimulus intensity required to produce this minimal neural response will vary greatly as stimulus frequency changes. As pointed out earlier, Güçlü and Bolanowski (2004) have determined that 5 to 10 RA fibers must be activated for stimulus detection to occur. Thus, it is our contention that a stimulus must produce a single spike in a minimum of 5 to 10 RA fibers at any given frequency for stimulus detection to occur.

The appropriate response criteria for constructing the tuning curves for the nerve fibers of the other three channels, however, are clearly different, as discussed earlier. Whatever the minimal neural response may actually be for each of the four channels, the principle established by this analysis is that the minimal neural response remains constant for the detection of stimuli at all frequencies. According to this principle, the minimal level of neural activity required for stimulus detection does not change as the frequency of the stimulus is changed, although the intensity of the stimulus may have to change substantially to produce the minimal level of neural activity necessary for detection to occur. In the tradition of Johannes Müller (1826), it is not the characteristics of the external stimulus but instead the specific neural response produced by the stimulus that determines the perceptual experience. In modern terms, our conclusion that constant neural responses in a tactile channel are equally detectable is an example of the *principle of nomination* (Marks, 1978), which states that identical neural events give rise to identical psychological events. Thus, according to this principle, when stimulus A and stimulus B produce the same minimal neural response required for stimulus detection within a tactile channel, they must produce the same sensory experience. The principle of nomination now evident in the tactile sensory system has been known for many years to operate in the visual system. Wald (1945) measured the rod (scotopic) sensitivity curve showing the physical intensities of stimuli of different wavelengths needed to produce identical sensations of barely detectable colorless light. According to the principle of nomination, these combinations of wavelengths and intensities of light must produce identical responses in the nervous system for identical sensory experiences to occur. Indeed, it has been shown that for the observer to detect light, the number of photons that must be absorbed by rhodopsin, the photochemical pigment in rods, is identical for all wavelengths, namely, 6 to 10 photons (for an analysis of this problem see Gescheider, 1997). According to Hecht, Shlaer, and Pirenne

(1942) this would require that a single photon be absorbed by one molecule of rhodopsin in each of 6 to 10 rods. In the tradition of Hecht and his co-workers, Güçlü and Bolanowski (2002, 2004) have found that 5 to 10 RA fibers must be activated for detection to occur in the RA channel. The number of fibers that must be activated for detection to occur in the other three tactile channels is yet to be determined.

We have argued that the frequency selectivity unique to each channel is determined at the receptor level. This argument is based on the fact that the frequency selectivity of the channel measured psychophysically is essentially identical in form to the frequency-tuning characteristic of the afferent nerve fibers of the receptors constituting the input stage to the channel. But what determines the frequency selectivity of afferent fibers? It has often been assumed that the tuning of a fiber is determined by the physical characteristics of the end organ that surrounds it. This assumption is based on the finding that the end organ responds selectively to stimulation. For example, the capsule of the Pacinian corpuscle surrounding the PC nerve fiber endows it with its rapidly adapting characteristic (Loewenstein & Mendelson, 1965). With the capsule intact, the PC fiber rapidly adapts to steady indentation of the corpuscle by a probe and exhibits only a few action potentials at the onset and offset of the stimulus, but none during continuous application of the probe. But when the capsule is surgically removed and the nerve fiber is deflected by the probe, the fiber slowly adapts and exhibits action potentials throughout the entire period of continuous stimulation. Thus, the capsule surrounding the PC nerve fiber causes the fiber to adapt rapidly, and therefore the fiber must be particularly responsive to a stimulus that rapidly changes in intensity, such as high-frequency vibration. This finding suggests that the capsule acts as a high-pass mechanical filter that attenuates the transmission of vibration to the nerve fiber to a greater extent at lower than at higher vibratory frequencies. If this is true, the frequency tuning of the PC nerve fiber must be at least partially determined by the mechanical tuning of the capsule with respect to the effectiveness with which it conducts the vibratory stimulus to the fiber it surrounds.

Loewenstein and Skalak (1966) developed a model of Pacinian-corpuscle mechanics in which the capsule functions as a high-pass filter for vibratory stimuli transmitted from the surface of the capsule through its lamellar structure to the core region of the corpuscle, where the unmylineated tip of its afferent nerve fiber resides. Because of the high-pass mechanical filtering by the capsule, high-frequency vibration is attenuated less than low-frequency vibration, and this must account, at least in part, for the greater sensitivity of the PC channel to high- than to low-frequency stimulation. Thus, according to this model, at least one component of the frequency tuning of the PC fiber is the selective mechanical filtering capacity of the Pacinian capsule surrounding it.

Although the capsule of the Pacinian corpuscle appears to function as a mechanical filter responsible for the frequency tuning of the PC nerve fiber, it has been suggested that neural characteristics of the nerve fiber itself also contribute to the frequency selectivity of the PC channel. The tuning of the PC fiber is affected by the temperature of the bathing solution in which an excised Pacinian corpuscle is placed (Bolanowski & Verrillo, 1982). As seen in Figure 3.5, maximum sensitivity shifts to a higher stimulus frequency as the temperature of the bathing solution increases. Bolanowski (1981) argued that this effect must be neural inasmuch as temperature has no significant effect on the mechanics of the capsule.

The details of how mechanical and neural factors influence the frequency tuning of PC nerve fibers remain under investigation, but at this time we can say that both factors may be important. Less is known about how mechanical and neural factors contribute to the tuning of RA, SA I, and SA II nerve fibers. The end organs surrounding these fibers are much smaller than Pacinian corpuscles and have therefore been much more difficult to study in ways that would reveal the sources of their frequency tuning.

Temporal Summation and Temporal Acuity

Temporal summation is operationally defined as a decrease in the detection threshold resulting from an increase in either stimulus duration or the number of successive brief stimuli rapidly presented to a sensory system. This phenomenon occurs in the sense of touch under very limited conditions. Indeed, it is only when high-frequency vibratory stimuli are delivered through a large contactor to an area of the skin containing Pacinian corpuscles that the detection threshold decreases as stimulus duration increases—a phenomenon that occurs only up to a stimulus duration of about 1 second (Gescheider, 1976; Gescheider, Beiles, Bolanowski, Checkosky, & Verrillo, 1994; Gescheider, Hoffman, Harrison, Travis, & Bolanowski, 1994; Gescheider, Zwislocki, & Rasmussen, 1996; Gescheider et al., 1999; Verrillo, 1965). From these results we conclude that the PC channel is the only one of the four tactile information-processing channels capable of temporal summation (see Figure 2.6). In contrast, thresholds are independent of stimulus duration when the same high-frequency stimuli are applied to the skin through a small contactor incapable of exciting the PC channel at the detection threshold but capable of exciting the SA II channel (Gescheider, 1976; Verrillo, 1965). At low vibration frequencies that excite only the RA channel, temporal summation is essentially absent (Gescheider, 1976; Green, 1976).

The decline in the detection threshold with increasing stimulus duration observed in the PC channel is predicted from Zwislocki's (1960) theory of temporal neural integration (e.g., Verrillo, 1965). In this model the magnitude of the neural response increases during stimulation, and when it exceeds the neural threshold, the observer makes a detection response. Because this neural threshold is reached sooner with more intense stimuli, the inverse relationship found between the psychophysical threshold and stimulus duration can be explained (Figure 5.2). According to the theory of neural integration, one must conceptualize an ongoing stimulus as consisting of separate brief stimulus events, each capable of causing an abrupt increase in neural activity followed by an exponential decay. It is this residual decay of persisting neural activity following each stimulus event comprising a stimulus summated over the total duration of the stimulus that results in the buildup of neural activity over the duration of the stimulus presentation. Under these conditions, presentation of a more intense stimulus of shorter duration produces larger neural responses that summate more rapidly to exceed the neural threshold than is the case when a weaker stimulus of longer duration is presented. Thus, according to this model, the psychophysical threshold expressed in terms of the intensity of a stimulus needed for detection should be higher for short- than for long-duration stimuli. This model can be tested in experiments on the detection of multiple-pulse stimuli.

According to the temporal neural-integration model, when brief stimulus pulses of equal intensity in a multiple-pulse stimulus are separated by short time intervals of a few milliseconds, there should be ample opportunity for the persisting neural response to the first pulse to add to the neural activity generated by a subsequent pulse. However, this should not be the case when the pulses are separated by much greater time intervals. If the threshold for detecting a multiple-pulse stimulus is exceeded when the neural response achieves a specific level, then the detection threshold expressed as the intensity of the pulses required for detection should decline as the time between pulses becomes shorter. Furthermore, according to the theory of temporal neural integration, the detection threshold should decrease as the number of pulses is increased,

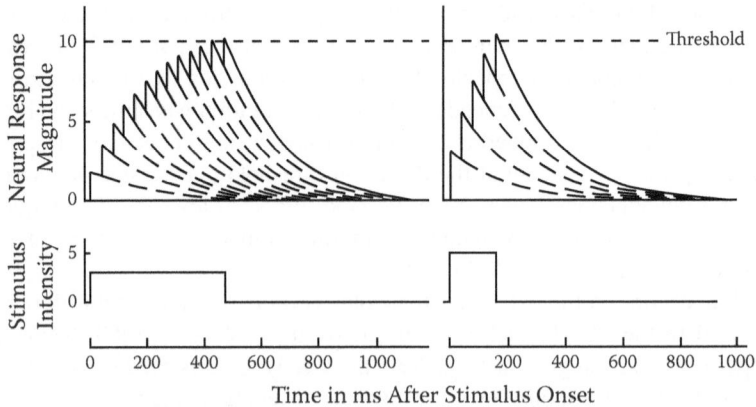

Figure 5.2 Illustration of Zwislocki's (1960) temporal-summation model applied to changes in neural responses as a function of the intensity and duration of a continuous stimulus.

but only when the time between pulses is sufficiently short to permit the occurrence of neural integration. These predictions have been confirmed for the PC channel (Gescheider et al., 1999). The results are shown in Figure 5.3, where it can be seen that thresholds for detecting multiple-pulse stimuli comprised of sequentially presented 10-ms pulses decrease as the number of pulses increases and the time between pulses becomes shorter. The predicted neural responses necessary to exceed the neural threshold and the corresponding stimulus pulses that produce them are shown in Figure 5.4. In this figure the neural

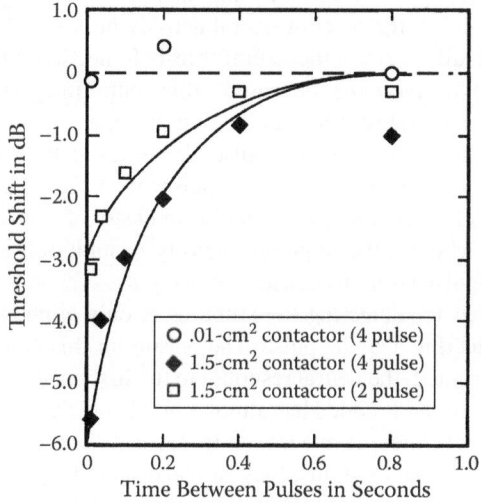

Figure 5.3 Threshold shift in dB relative to the threshold for detecting a single pulse as a function of the time between pulses and the number of pulses in a multipulse stimulus. The curves are predicted from Zwislocki's (1960) neural-integration model. Data are from Gescheider et al. (1999).

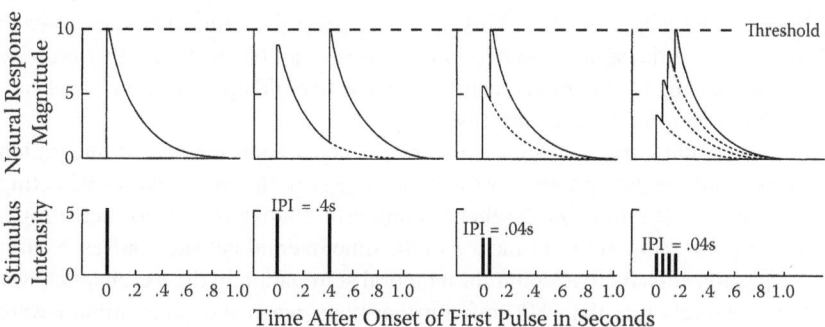

Figure 5.4 Illustration of Zwislocki's (1960) theory of neural integration as applied to the detection of single-pulse and multipulse stimuli.

responses are graphically illustrated, each with its abrupt onset in response to the onset of the pulse and exponential decay following the termination of the pulse. To facilitate understanding of Zwislocki's neural-integration model of temporal summation as it applies to the detection of brief tactile stimuli, neural responses to the following stimulus situations are presented in the figure: presentation of a single pulse; presentation of two pulses separated by 400 ms; presentation of two pulses separated by 40 ms; and presentation of four pulses, each separated by 40 ms. It is evident that because the neural response to an individual stimulus pulse adds to the exponentially decaying neural response to the previous pulse, the level of neural activity builds with each successive pulse and eventually reaches the neural threshold needed for detection of the multipulse stimulus. Note that because of this neural integration, the required magnitude of the neural response to the first pulse, as well as the incremental neural response to each successive pulse, decreases as the interpulse interval decreases and the number of successive pulses increases. Thus, to exceed the neural threshold for detection, the stimulus intensity of a single pulse must be relatively high, whereas the required intensity of multipulse stimuli is lower. When the multipulse stimulus consists of two pulses separated by 400 ms, the intensity required for detecting the stimulus is only slightly lower than that required for detecting a single pulse. The reason for this is that when the second pulse is presented, the neural response to the first pulse to which the neural response of the second is added has almost completely decayed, thus providing little opportunity for neural integration to occur. When the time between the two pulses is shortened to 40 ms, they can be detected at a substantially lower intensity level than is required when separated by 400 ms. According to the neural-integration model, this is because the neural response to the first pulse has decayed very little when the second pulse is presented, thus resulting in greater opportunity for neural integration to occur. Finally, it can be seen that the intensity required to detect a multipulse stimulus consisting of four pulses separated by 40 ms is even lower than that for detecting two pulses separated by the same length of time. This occurs because the neural response to each pulse adds to the summated neural responses of all of the pulses that precede it, thus causing the integrated neural response to build up as the number of successively presented pulses increases.

The solid curves of Figure 5.3 are predicted from the neural-integration model, and the data points represent the change in the threshold for detecting a multipulse stimulus on the thenar eminence relative to that for detecting a single-pulse stimulus as a function of the time interval between pulses. Neural integration appears to account for temporal summation in the PC channel over time intervals less than 800 ms. Furthermore, when the same stimuli were delivered through a very small contactor, threshold was independent of the number of pulses presented and the time interval between them. This indicates that neither temporal summation nor neural integration occurs in the SA II channel, which is responsible for the detection of high-frequency stimuli presented through a small contactor (Gescheider et al., 1999).

But where in the nervous system does the summation of these neural responses occur? It is our contention that summation must occur in neurons within the central nervous system that continue to be active for as long as 800–1000 ms following the termination of a stimulus. This hypothesis is based on the following two facts: (1) temporal summation measured psychophysically is possible for time intervals as great as 800–1000 ms; and (2) there is essentially no persisting neural activity in peripheral afferent PC fibers following the termination of a vibratory stimulus (Bolanowski, 1981). Hence, by default, neural integration must occur within the central nervous system.

The finding that temporal summation is based on neural integration and is a unique property of the PC channel provides an opportunity to examine the relationship between temporal summation and temporal acuity. A sensory system capable of temporal summation resulting from the neural persistence required for summation may in fact be poor at temporal acuity. This is because persisting neural activity in such a system may fill the temporal gap between the termination of one stimulus and the onset of the next, thus rendering the gap undetectable, so that instead of being perceived as separate, successive stimuli are incorrectly perceived as one continuous stimulus. This is evident in vision, in which neural persistence is easily demonstrated through experiments in critical flicker fusion (Kelly, 1961), but temporal resolution is poor relative to that observed in hearing where neural persistence is minimal (Viemeister & Wakefield, 1991). Therefore, one possible hypothesis to account for temporal summation in the PC channel and the lack thereof in the other channels is that there is greater neural persistence in the PC channel. If this is true, then temporal acuity as measured by temporal gap-detection thresholds should be poorer in the PC channel than in the other channels.

To measure gap-detection thresholds in the PC channel, a 250-Hz stimulus was applied to the thenar eminence through a relatively large 1.5-cm^2 contactor. To measure gap detection in the SA II channel, the same 250-Hz stimulus was applied through a small 0.01-cm^2 contactor; and to measure gap detection in the RA channel, a much lower frequency 62-Hz stimulus was applied through the small 0.01-cm^2 contactor (Gescheider, Bolanowski, & Chatterton, 2003). Plotted in Figure 5.5 are gap-detection thresholds expressed as the duration of an interruption in an ongoing vibratory stimulus necessary for the observer to detect the interruption. It can be seen that the gap-detection threshold declines rapidly as the intensity of the stimulus increases. Presumably, the improvement in temporal acuity with increases in stimulus intensity occurs because of the greater change in neural activity level resulting from the interruption of a strong stimulus rather than from the interruption of a weak stimulus. Most important for the present discussion, however, is the finding of no significant differences among the gap-detection thresholds for the PC, SA

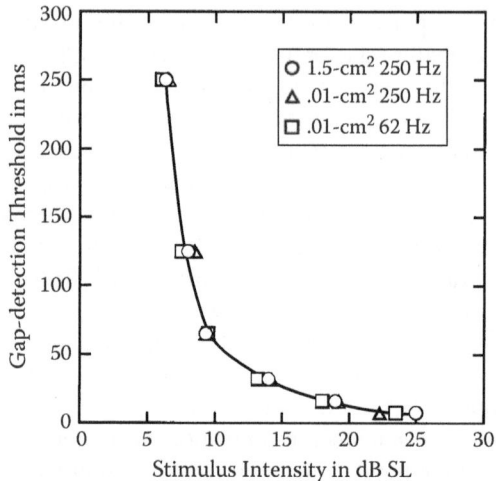

Figure 5.5   Gap-detection thresholds in ms as a function of stimulus intensity in dB SL (decibels above the detection threshold) for stimuli chosen to preferentially activate the PC channel (1.5-cm² contactor, 250 Hz), the SA II channel (0.01-cm² contactor, 250 Hz), and the RA channel (0.01-cm² contactor, 62 Hz). Data from Gescheider et al. (2003).

II, and RA channels at any stimulus-intensity level. This strongly suggests that temporal acuity does not differ among tactile channels. This hypothesis is also supported by the finding that temporal modulation transfer functions, as measures of temporal acuity, do not differ for low- and high-frequency stimuli that selectively activate RA and PC channels, respectively (Weisenberger, 1986). Therefore, the lack of temporal summation in the SA II and RA channels must result not from a lack of neural persistence in these channels but instead from the absence of a neural mechanism capable of integrating persisting neural activity. If neural persistence in these channels had been substantially less than in the PC channel, then they would have been superior to the PC channel in their capacity to resolve temporal gaps in stimulation. Inasmuch as the channels do not differ in their ability to detect gaps in ongoing stimulation, it follows that they do not differ in how much neural activity persists following the termination of a stimulus.

Several findings can now be explained by a model in which neural persistence is presumed to be essentially the same in the temporally summating PC channel as in the non-summating tactile channels. First is the finding seen in Figure 4.1B (Verrillo & Gescheider, 1975) that the phenomenon of enhancement occurs to the same degree when low-frequency stimuli predominantly excite non-Pacinian channels (25-25-25 Hz) as when high-frequency stimuli exclusively excite the PC channel (300-300-300 Hz) (Verrillo & Gescheider, 1975). The enhancement of the sensation magnitude of a brief 20-ms test stimulus by prior presentation of a conditioning stimulus of higher intensity has been explained as a process in which the persisting neural activity of the conditioning stimulus adds to the neural activity generated by the subsequently presented test stimulus, thus rendering it perceptually more intense. The finding that the amount of enhancement measured experimentally decreases as the time between the conditioning stimulus and the test stimulus increases is also consistent with this model. A model in which it is assumed that neural persistence is the same in the different tactile channels can explain the fact that the amount of enhancement measured at a particular time interval between stimuli is independent of stimulus frequency.

Second, the finding that the amount of forward masking measured at a particular time interval between a test stimulus and a preceding masking stimulus is not greater in the PC channel than in the RA channel (Gescheider, Bolanowski, & Verrillo, 1989; Gescheider, Santoro, Makous, & Bolanowski, 1995; Gescheider, Valetutti, Padula, & Verrillo, 1992; Gescheider & Migel, 1995; Makous, Gescheider, & Bolanowski, 1996) can be explained by a model in which it is assumed that the persisting neural activity that follows the termination of the masking stimulus and subsequently masks the detection of the test stimulus is not greater in the PC channel than in the other channels. The results of an experiment on forward masking that involved the masking of test stimuli presented through a large 1.5-cm^2 contactor to the thenar eminence are seen in Figure 5.6 (Gescheider et al., 1992). It can be seen that shifts in threshold for detecting a brief 50-ms stimulus presented at variable times following the termination of a 500-ms masking stimulus of the same frequency at an intensity level 20 dB above the detection threshold were not significantly different for the PC (250 Hz) and RA (25 Hz) channels.

Third, psychophysical measurements of how long sensations last following the termination of a vibratory stimulus are not affected by stimulus frequency (Gescheider et al., 1992). Again, a model in which the same amount of neural persistence following termination of stimulation is assumed to exist in the different tactile channels can explain these results. In one experiment, the observer adjusted the temporal onset of an auditory click so that it was perceived to match the onset or offset of a tactile stimulus applied to the thenar eminence. The results, shown in Figure 5.7, revealed that the click was set approximately 10 ms after the onset of the tactile stimulus, but 75–80 ms after its offset. The latter finding indicates that the tactile sensation substantially outlasted the stimulus—an effect attributable to neural persistence in the tactile system. Of particular importance for the present discussion is the fact that the duration of these tactile after-sensations did not differ between the PC (250 Hz) and RA (25 Hz) channels. This finding is consistent with a model in which the amount of neural persistence in the PC and RA channels is assumed

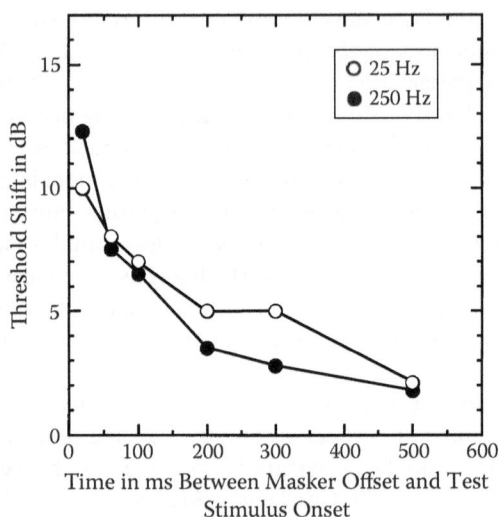

Figure 5.6 Forward masking expressed as a shift in the detection threshold attributable to presentation of the masking stimulus when stimuli preferentially activate the RA channel (25 Hz) and when they exclusively activate the PC channel (250 Hz). Data from Gescheider et al. (1992).

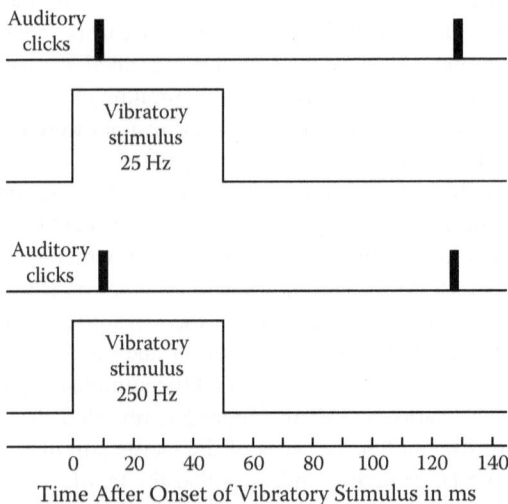

Figure 5.7 Measurement of the perceived duration of a 50-ms vibratory stimulus that preferentially activates the RA channel (25 Hz) or exclusively activates the PC channel (250 Hz).

not to differ, thus resulting in after-sensations in the PC and RA channels of equivalent duration.

In summary, measures of temporal acuity, enhancement, forward masking, and the duration of after-sensations support the hypothesis that the amount of neural persistence following the termination of a stimulus is the same in all tactile channels.

But why is temporal summation an exclusive property of the PC channel if temporal summation depends upon the integration of persisting neural activity? A consideration of the relationship between temporal summation and spatial summation helps to answer this question. Indeed, not only is the PC channel the only channel capable of temporal summation, it is also the only channel that exhibits spatial summation in which sensitivity improves as the size of the stimulus is increased (Bolanowski et al., 1988; Gescheider, 1976; Verrillo, 1963). Further, Checkosky and Bolanowski (1992) have found that the form of the function describing the decrease in the detection threshold of the PC channel as stimulus duration increases cannot be predicted from a model in which the neural activity of a single Pacinian corpuscle provides the necessary input to a theoretical temporal integrating mechanism assumed to be located in the central nervous system. Instead, it appears that the activity of several Pacinian corpuscles is required for temporal summation to occur.

This suggests that the operation of the *temporal* integration mechanism may be dependent upon the operation of a *spatial* integration mechanism for temporal summation to occur in the PC channel. Inasmuch as there is no evidence for the operation of either spatial or temporal integrating mechanisms in channels other than the PC channel, it is not surprising that both temporal and spatial summation are exclusive properties of the PC channel. Thus, although a sufficient amount of neural persistence occurs in all of the tactile channels for temporal summation to occur, it appears that only in the PC channel does a mechanism exist—one possibly associated with the mechanism responsible for spatial summation—that is capable of integrating neural activity over time. Whether temporal integration and spatial integration consist of two aspects of the same mechanism or exist as separate but interconnected mechanisms is a question yet to be answered. Nevertheless, what can be stated is that temporal summation cannot occur in the absence of spatial summation, and therefore temporal summation can occur only after some amount of spatial summation has taken place. This suggests that the neural mechanism responsible for temporal summation must be located at a site in the central nervous system higher than the site where spatial summation occurs.

Spatial Acuity

Do tactile channels differ in spatial acuity—the ability to resolve the small spatial details of objects that touch the skin? To answer this question, one must consider both the physiological and psychophysical characteristics of the channels as they pertain to spatial discrimination. There are two physiological characteristics of receptors that potentially affect spatial acuity: the density of the receptors and the size of their receptive fields.

It is well established that individual SA I and RA afferent fibers innervating Merkel and Meissner receptors, respectively, because of their very small receptive fields (Johansson, 1976), are exceptionally well suited for resolving small spatial details (Connor, Hsiao, Phillips, & Johnson, 1990). For example, as a small dot less than a millimeter in width and height rising from a smooth background is moved a few millimeters from the center of an SA I or RA fiber's receptive field to a point on the skin outside the receptive field, the firing rate of the fiber plummets from high to very low. The fact that the firing rate of the fiber can be high when the dot is within a small area of a few square millimeters and low when it is outside this area could potentially provide very precise information about the location of the dot on the skin. In contrast, individual PC and SA II afferent fibers innervating Pacinian corpuscles and SA II end organs, respectively, have exceptionally large receptive fields, often many square centimeters in area (see Figure 3.1), thus rendering them potentially poor at discriminating fine spatial details. A raised dot may have to be moved several centimeters, for example, to exit the receptive field of a PC fiber. Thus, neural activity in an individual PC fiber provides highly inaccurate information about the precise location of the dot on the skin.

From this observation, it has often been concluded that Pacinian corpuscles do not contribute significantly to spatial acuity in the sense of touch. However, this conclusion can be challenged on both empirical and theoretical grounds. Specifically, in spite of the large receptive fields of individual PC fibers, observers can detect the location of a vibratory stimulus applied to the hypothenar eminence (located below the little finger) almost as well when stimuli are presented to the PC channel (250 Hz) as when presented to the RA and SA I channels (25 Hz) (Sherrick, Cholewiak, & Collins, 1990). Shown in Figure 5.8 are the percentages of correct responses in identifying which of two contactors on the hypothenar—the distal or the proximal—was activated on a particular stimulus-presentation trial. Although performance was significantly better for 25-Hz stimuli (RA and SA I channels) than for 250-Hz stimuli (PC channel), the difference in the spatial acuity between the channels was small. When the positions of the contactors were moved to a more distal location on the hypothenar, where receptor density was higher, the channels did not differ in their performance on the spatial-localization task (Sherrick et al., 1990). Thus, it appears that with a sufficient density of receptors, spatial acuity in a localization task becomes independent of the size of the receptive fields of the nerve fibers innervating these receptors. Rogers (1970) demonstrated that two-point acuity on the fingertip, an area of even higher receptor density than the hypothenar, was actually slightly better for 250-Hz stimuli (PC channel) than

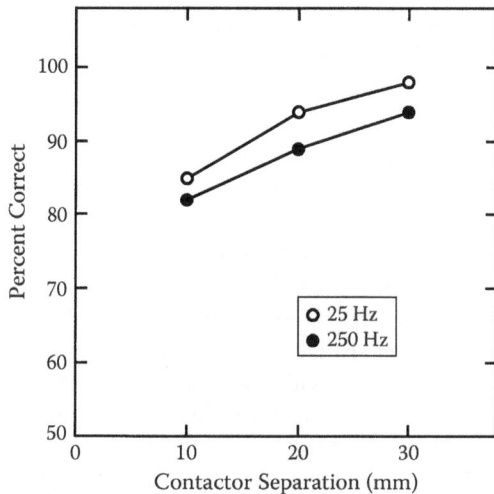

Figure 5.8 Percentage of correct localizations as a function of contactor separation in a two-alternative forced-choice task in which the observer indicates which of two spatially separated contactors was activated. From Sherrick et al. (1990).

for 10-Hz stimuli (RA and SA I channels). Sherrick et al. (1990) proposed that good spatial resolution in the PC channel occurs when the pattern of activity in a number of PC nerve fibers with overlapping receptive fields and broad spatial tuning changes systematically as the spatial characteristics of the stimulus change. Based on observations of the activity of individual PC fibers, it is clearly incorrect to conclude that the PC channel, consisting of a population of PC fibers with their central connections, is incapable of conveying substantial amounts of spatial information.

The assumption that the spatial resolving power of individual tactile nerve fibers determines tactile spatial acuity measured psychophysically appears to have its origin in the *labeled-line* theory of sensory coding, which evolved from Müller's (1826) doctrine of specific nerve energies. According to this theory, each sensory fiber provides a *line* with its own *labeled* meaning to the central nervous system. For example, there are neurons in vision that when activated signal "red," and others that signal "green." In the sense of touch there are neurons that signal that a specific spot on the skin has been stimulated. The limitations of the labeled-line theory have been known for many years (Erickson, 1968, 1978, 1982). Simply put, the individual neurons of sensory receptors are generally broadly tuned relative to the discriminative capacities of the organism, and therefore the activity of any single neuron must be considered within the context of the activity of the other neurons within the surrounding population of neurons. For example, in color vision the activity of a "green" sensitive cone has meaning in determining the perceived color (green, yellow, blue-green, etc.) only within the context of the activity of "blue" and "red" sensitive cones, as the Young-Helmholtz theory of color vision states (Erickson, 1982). This general explanatory concept, referred to as *across-fiber pattern theory* (see Erickson, 1982), is also applicable to gustatory (Erickson & Schiffman, 1975) and olfactory (Buck, 1996; Buck & Axel, 1991) sensation, in which activity patterns across a relatively small number of specific receptor types encode an enormous number of tastes and smells.

The work of Sherrick et al. (1990) and Rogers (1970) shows that the across-fiber principle of sensory coding can be applied to the sense of touch. Specifically, the excellent spatial acuity of the PC channel can be explained by a pattern of neural activity within a *population* of PC afferent fibers. This concept is illustrated in Figure 5.9 for a situation in which two tactile stimuli separated by a distance substantially smaller than the size of the receptive field of any single PC fiber can nevertheless be discriminated as different by the pattern of neural activity in several broadly tuned PC fibers with overlapping receptive fields. The hypothetical response function for each PC fiber shows

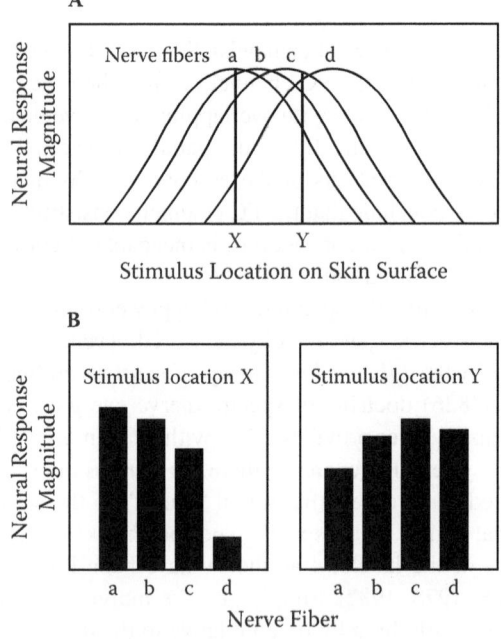

Figure 5.9 Hypothetical example of the responses within the receptive fields of individual PC fibers a, b, c, and d (A) and the resulting pattern of neural activity across fibers a, b, c, and d for two spatially separated stimuli applied to the skin (B).

very broad spatial tuning, with the highest response magnitude occurring in the middle of the receptive field, and the response magnitude declining as the distance from the center increases (Figure 5.9A). The receptive fields of PC fibers overlap sufficiently so that each of the four fibers in this example is activated by the presentation of stimulus X or stimulus Y. It can be seen that the resulting response profiles across the four fibers generated by separately applying each of the two stimuli are very different (Figure 5.9B), and this could well provide the neural basis for perceiving the difference in their location on the skin.

The formulation of a more precise across-fiber model capable of generating predictions of an observer's psychophysical performance in spatial-acuity tasks awaits the following information: (1) a detailed description of the spatial distribution of tactile receptors within a particular area of skin; (2) quantitative specification of the response function of the receptive fields of individual nerve fibers; and (3) extensive psychophysical measurements of performance in spatial-acuity tasks under conditions known to generate specific changes in the patterns of activity across these nerve fibers. Until this information becomes available, we are limited in our ability to link patterns of spatially distributed neural activity in tactile nerve fibers to psychophysically measured spatial-discrimination performance. But at this time we can say with confidence that the relatively high level of spatial acuity found in the PC channel containing afferent fibers with large receptive fields can be explained by the principle of across-fiber coding.

Edge Detection

In experiments on spatial summation, the relationship between the threshold for detecting the stimulus and the size of the skin area to which it is applied is examined. For this to be accomplished successfully, it is essential that the stimulus be confined to the area of skin directly under the contactor as contactor size is systematically varied. To accomplish this, a rigid surround (see Figure 3.4) is used to prevent the spread of surface waves originating from the vibratory stimulus to areas of the skin beyond the area directly under the contactor. Thus, by using the surround, the size of skin area stimulated is essentially the same as the size of the contactor. As described earlier, increasing the size of the contactor under these conditions lowers the detection threshold at high but not at low stimulus frequencies, thus demonstrating that spatial summation occurs in the PC channel but not in the channels mediating the detection of low-frequency stimuli. In contrast, when the size of the contactor is increased in the absence of a rigid surround, not only are the effects of contactor size on threshold greatly reduced in the PC channel, but thresholds for detecting low-frequency stimuli by the other channels actually increase (Gescheider et al., 1978; Verrillo, 1962; Verrillo, 1979b). Two important changes in the stimulus occur when the rigid surround is removed: (1) the effective area of stimulation is increased as vibration is permitted to spread along the surface of the skin; and (2) the sharp gradient between the skin area stimulated and the skin area not stimulated is eliminated.

The effects on sensitivity of the spread of the stimulus beyond the contactor and the corresponding reduction in the sharpness of the spatial gradient in stimulation can be examined by measuring the detection threshold as the distance between the edge of the contactor and the edge of the rigid surround is increased (Figure 5.10). Progressively increasing the distance between the contactor and the surround increases the effective area of stimulation on the skin, but at the same time decreases the slope of the spatial gradient of stimulation. As seen in Figure 5.11, as the gap between a 0.32-cm² contactor and surround is increased, thus allowing the vibratory stimulus to spread over a larger area, detection thresholds at 250 Hz decrease as a result of spatial summation in the PC channel. But because of the lack of spatial summation in the

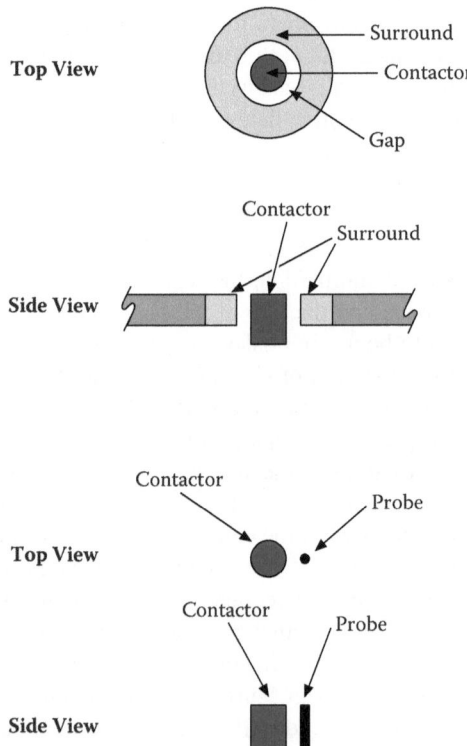

Figure 5.10 Top and side views of contactor, surround, and probe configurations used by Verrillo (1979b).

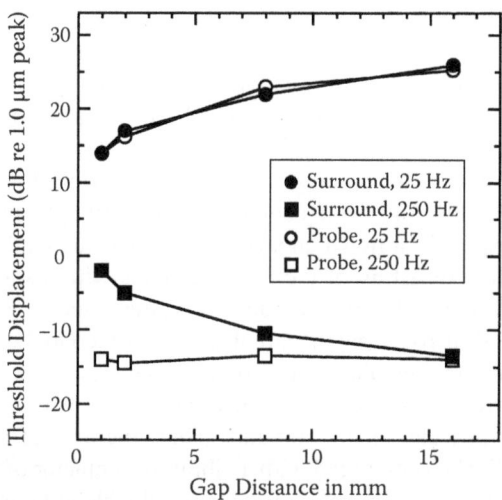

Figure 5.11 Thresholds for detection of 25-Hz and 250-Hz stimuli plotted as a function of gap distance. The gap was either the distance in millimeters between the edge of the contactor and the edge of a rigid surround or, in the absence of a surround, the distance between the edge of the contactor and a small, single, nonvibrating probe. From Verrillo (1979b).

channels that mediate the detection of low-frequency stimuli (RA and SA I channels), no such reductions in thresholds are observed at 25 Hz. Instead, the threshold for detecting the 25-Hz stimulus actually increases as the gap between the contactor and surround increases. This indicates that RA and/or SA I receptors are stimulated more effectively when the gap is small, which causes the spatial gradient between the area of stimulation and non-stimulation to be steep. Essentially the same results were obtained when the surround was removed and replaced by a small 0.005-cm² nonvibrating probe. As the distance between the probe and the contactor edge was systematically increased, the same loss of sensitivity for detecting the 25-Hz stimulus was observed (Figure 5.11). Thus, a single small but well defined edge produced by a stationary probe located close to the contactor in an essentially edgeless field of vibration substantially improves the sensitivity of the non-PC channels. The results for the 250-Hz stimulus, on the other hand, were quite different for the probe than for the rigid surround. Whereas increasing the distance of the probe from the contactor edge had no effect upon the threshold for detecting a 250-Hz stimulus, increasing the distance of the rigid surround from the contactor edge lowered the 250-Hz detection threshold appreciably because of greater opportunity for spatial summation to occur (Verrillo, 1979b).

It is important to note that the PC and non-PC channels behave quite differently as the distance of either the rigid surround or the stationary probe from the contactor edge is varied, again showing that the channels operate in fundamentally different ways in detecting tactile stimuli. The PC channel is uniquely capable of spatial summation, whereas the low-frequency non-PC channels are highly effective in detecting sharp spatial gradients in stimulation, namely *edges*.

The hypothesis that spatial gradients are important for stimulating the receptors of the RA and SA I channels is also strongly supported by the results of an experiment in which a rigid surround was used in combination with a contactor consisting of a circular annulus with a solid nonvibrating core within (Verrillo, 1963). This device is illustrated in Figure 5.12. The areas within and surrounding the annulus were affixed to the rigid surface of a table with only the annulus mounted on the vibrator. When the observer's hand was in position, the skin was free to move only within an area limited to 1 mm on either side of the annulus. Thus, the area of stimulation was relatively small when compared with a vibrating contactor of the same diameter, but the pattern of stimulation produced a double gradient. One gradient was produced between the outer edge of the annulus and the rigid surround, and the second gradient was located between the inner edge of the annulus and the solid nonvibrating core. Thresholds measured over a wide range of stimulus frequencies with the annulus and with a standard contactor of equal diameter are shown in Figure 5.13. Use of the annulus resulted in higher thresholds than did the standard contactor for the high-frequency stimuli necessary for PC-channel activation. Sensitivity of the spatially summating PC channel was clearly compromised by the reduction in the area of stimulation resulting from use of the annulus. At low frequencies, on the other hand, thresholds are determined by NP receptors, and the double gradient of stimulation produced by the annulus enhanced their sensitivity and consequently lowered their detection thresholds. Thus, it can be seen that the PC channel is relatively unresponsive to changes in the spatial gradient of a stimulus, whereas the NP channels that detect low-frequency vibratory stimuli clearly are responsive to changes in the spatial gradient.

Of the two NP channels that detect low-frequency stimuli, the SA I channel appears to be particularly sensitive to the steep spatial gradient created when the edge of an object makes contact with the skin (Goodwin, Browning, & Wheat, 1995; Johansson, Landström, & Lundström, 1982b; Phillips & Johnson, 1981; Vierck, 1979). Placing a finger on the edge of an object, for example, results in spike rates in SA I fibers that are 20 times higher than when placing the finger on a smooth surface (Phillips & Johnson, 1981). Furthermore, both RA and SA I fibers in human glabrous skin have been found to be much more responsive to vibratory stimuli when the edge of the contactor is placed within the receptive field of the fiber than when the receptive field is completely cov-

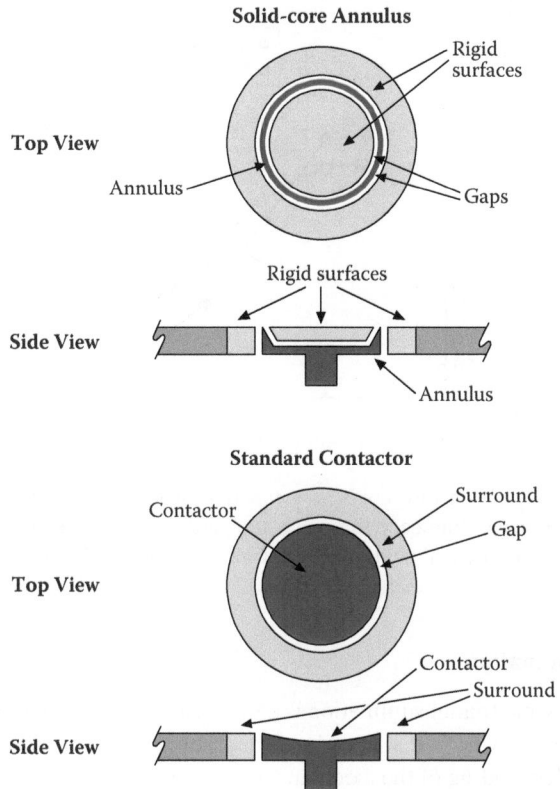

Figure 5.12   Top and side views of standard contactor and solid-core annulus.

ered by the contactor (Johansson et al., 1982b). This effect was found to be greater for SA I than for RA fibers. Thus, the SA I channel appears to be particularly well suited for detecting the elements of a textured surface defined by edges (Connor & Johnson, 1992; Johnson & Hsiao, 1992; Johnson, Yoshioka, & Vega-Bermudez, 2000). But it should not be concluded from these findings that texture perception results exclusively from activity in the SA I channel. The involvement of the PC channel in the perception of the roughness of a surface is now well established (Bolanowski, Cohen, & Pawson, 1997; Gescheider, Bolanowski, Greenfield, & Brunette, 2005; Hollins, Bensmaia, & Risner, 1998; Hollins, Bensmaia, & Washburn, 2001; Hollins, Fox, & Bishop, 2000; Hollins & Risner, 2000). The results of these studies show clearly that the PC channel responds differently to spatial gradients than do the SA I and RA channels, which constitutes further evidence that the channels operate separately and independently in responding to tactile stimulation.

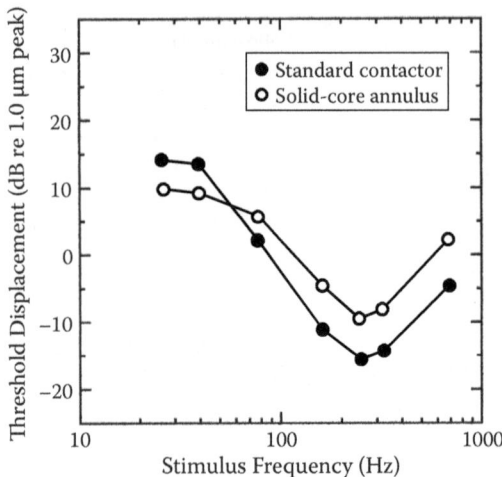

Figure 5.13   Detection thresholds as a function of stimulus frequency for stimuli applied to the thenar eminence through a standard 2.9-cm² contactor and through a rigid-core annulus of equal circumference. From Verrillo (1963).

Spatial Summation

As we have seen, spatial summation is an exclusive property of the PC channel, (Figure 2.5). But why is it seen only in this channel and not in the others? Perhaps an understanding of the mechanisms that underlie the inverse relationship that exists between the size of the contactor through which the stimulus is delivered and the detection threshold mediated by Pacinian corpuscles can explain this finding. As the size of the contactor increases, more Pacinian corpuscles are exposed to the stimulus, and at least two things can potentially happen as a result. First, neural activity originating from those receptors activated by the stimulus can possibly summate their neural responses at some point in the central nervous system—a process referred to as *neural integration*. Second, because the neural thresholds of individual Pacinian corpuscles are distributed over a wide range of intensities (Bolanowski, 1981), the likelihood that the most sensitive receptors will be exposed to the stimulus increases—a process called *probability summation*. The operation of either or both of these processes could cause the decrease in the psychophysical detection threshold observed as the size of the contactor delivering the stimulus to the receptors increases.

There is psychophysical evidence that both neural integration and probability summation operate in spatial summation in the PC channel (Gescheider, Güçlü, Sexton, Karalunas, & Fontana, 2005; Güçlü, Gescheider, Bolanowski, & Istefanopulos, 2005). Cumulative frequency distributions of thresholds for the detection of a 250-Hz stimulus applied through a large 1.5-cm² contactor and through a small 0.01-cm² contactor are shown in Figure 5.14. The thresholds were obtained from randomly sampled sites within a 3-cm² area on the thenar eminence. If spatial summation is accounted for entirely by probability summation, then the lowest detection thresholds sampled within the 3-cm² testing area with the large and small contactors should not differ. This is because with both contactors, the most sensitive receptors with the lowest neural thresholds would determine the psychophysical threshold. In contrast, if neural integration alone accounts for spatial summation, then the cumulative frequency distributions of thresholds obtained with the large and small contactors should have the same slope, but with the distribution of thresholds for the large contactor being located at lower stimulus intensities than those for the small contactor. Such a finding would indicate that the difference in thresholds measured with the two contactors does not result at all from probability summation and therefore must result from another process—presumably neural integration. The results confirm neither of these predictions. Instead, they strongly suggest that both probability summation and neural integration operate to produce spatial summation in the PC channel. Examination of the cumulative distributions of thresholds for the two contactor sizes indicates that the average difference in thresholds at the 0.5 point of the two distributions is

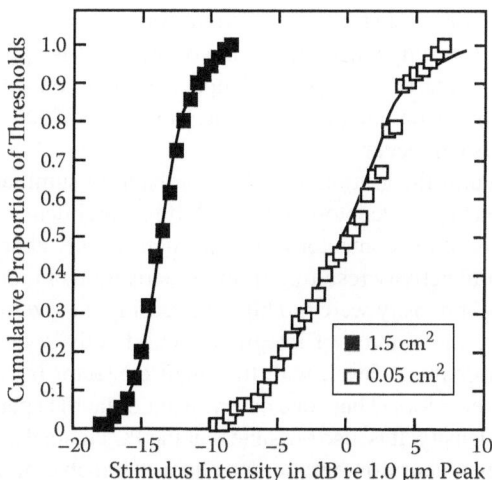

Figure 5.14 Cumulative probability distributions of detection thresholds for small (0.05-cm²) and large (1.5-cm²) contactors with the spatial position of the contactor varied within a 3-cm² area on the thenar eminence. The solid curves are best-fitting cumulative Gaussian functions. From Gescheider, Güçlü, Sexton, Karalunas, and Fontana (2005).

13 dB, whereas it is only 8 dB at the lowest thresholds of the distributions. If spatial summation were due entirely to neural integration, then the difference between the thresholds measured with the two contactors would have been 13 dB at the average thresholds and also at the lowest thresholds. If probability summation had been operating alone, then the lowest thresholds would not have differed for the two contactors, whereas the average threshold could have differed considerably. Thus, a mechanism other than probability summation, presumably neural integration, must account for the 8-dB difference in the lowest thresholds obtained with the small and large contactors.

Why then is spatial summation observed only in the PC channel? The answer may have to do, in part, with the density of receptors in the skin (see Gescheider et al., 2002). Probability summation requires that: (1) the neural thresholds of receptors of a particular type must differ appreciably; and (2) the density of these receptors must be optimal for the probability of stimulation of the most sensitive of these receptors to increase as the size of the contactor is increased. If receptor density were very low, then increasing the size of the contactor would not significantly change the number of receptors exposed to the stimulus; as a result, the probability of activating the most sensitive receptors would change very little, and spatial summation resulting from probability summation would be negligible. If receptor density were very high, then the number of receptors exposed to the stimulus would be high for both small and large contactors. Consequently, the probability of activating the most sensitive receptors would be high for both large and small contactors, and spatial summation resulting from probability summation would therefore be negligible in this case as well. It is only when the density of receptors is optimal in relation to changes in the size of the stimulus applied by the contactor that spatial summation resulting from an increase in the probability of activating the most sensitive receptors can occur.

Likewise, neural integration as well as probability summation can occur only when the density of receptors is optimal. If receptor density were too low, increasing the size of the contactor would not significantly change the amount of integrated neural activity resulting from the activation of more receptors. In contrast, if receptor density were too high, increasing the size of the contactor would not change the amount of integrated neural activity because it would already be at a maximum level, with the small contactor having activated a large number of receptors. Thus, one determinant of tactile spatial summation may be receptor density. It is also possible that the PC channel is the only channel that possesses a neural-integration mechanism capable of summating the neural activity that originates in individual receptors.

The exact location within the nervous system where neural integration over space occurs is not yet known, but we can rule out peripheral nerve fibers, central nervous system neurons in the spinal cord, and their target site in the brainstem—the dorsal column nuclei. Sensory neural integration requires that sensory afferent neurons synaptically interact with neurons higher in the nervous system so that the activity level in these higher neurons can increase as the number of afferent neurons providing input to them increases. Thus, neural integration cannot occur in PC afferent fibers because they do not synapse with each other in the peripheral nervous system. The first site where PC fibers synapse is in the brainstem, where they synapse with neurons in the dorsal column nuclei. But this appears not to be the site of neural integration because it has been found that a single neural impulse in one sensory fiber can generate activity in single dorsal column neurons (Rowe, 2002). The implication of this finding is that neural integration, in which several active neurons at a lower level cause activity in a neuron at a higher level, must occur at a higher level in the central nervous system than the dorsal column nuclei.

The same reasoning applies to the location of the site for neural integration over time demonstrated to occur in temporal summation (Gescheider et al., 1999). Temporal integration must occur at some site central to the dorsal column nuclei as well. Recall that Checkosky and Bolanowski (1992) demonstrated that the neural activity provided by a single PC fiber as input to a temporal neural integrator cannot account for psychophysically measured temporal summation in the PC channel. If more than one PC fiber must be active to account for temporal summation, then these fibers must interact. We now know from Rowe's (2002) work that this interaction must occur central to the dorsal column nuclei. Thus, temporal summation requiring neural integration over both space and time must occur at one or more sites central to the dorsal column nuclei.

Effects of Observer Characteristics

Effects of Aging on the Sensitivity of Tactile Channels

The sense of touch, like other sensory systems, suffers from the deleterious effects of aging (Frisina & Gescheider, 1977; Kenshalo, 1979; Verrillo, 1979a, 1982). Perhaps the most interesting aspect of this finding is that the rate at which tactile sensitivity declines with age is not the same for each of the tactile channels. This is clearly shown in the results presented in Figure 5.15, in which the average detection threshold on the thenar eminence is plotted as a function of the average age of eight groups of observers (Gescheider, Bolanowski, Hall, Hoffman, & Verrillo, 1994).

The sensitivities of the PC, RA, and SA I channels were assessed by measuring thresholds for detecting 250-Hz, 10-Hz, and 1-Hz stimuli, respectively, all delivered through a 2.9-cm^2 contactor, and the thresholds of the SA II channel were measured by presenting the 250-Hz stimuli through a 0.008-cm^2 contactor. By doing this, the effects of aging on the sensitivity of each of the four channels could be independently determined.

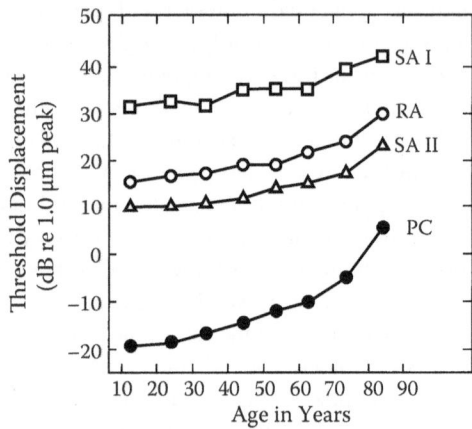

Figure 5.15   Detection thresholds of the SA I, RA, SA II, and PC channels as a function of age. Data from Gescheider, Bolanowski, Hall, Hoffman, and Verrillo (1994).

The effects of aging were substantially greater in the PC channel than in the other three channels, each of which showed upward shifts in threshold of 10 to 12 dB over approximately eight decades. A much larger upward shift of 24 dB was seen in the PC channel. Why should the effects of aging be greater in the PC than in the RA, SA I, or SA II channels? Part of the answer may be found in the fact that the PC channel, but none of the other channels, is capable of spatial summation. As shown in Figure 2.5, stimulating more tactile receptors by increasing the size of the contactor greatly lowers the detection threshold, but only in the PC channel. The progressive loss of receptors with age (Cauna, 1965) should greatly reduce the sensitivity of the spatially-summating PC channel inasmuch as the exceptional sensitivity of this channel depends to a large degree on the number of activated receptors. Much smaller effects of aging are expected in the other channels because studies of spatial summation indicate that their sensitivities are nearly independent of the number of receptors activated. Consequently, the loss of receptors with aging has a much smaller effect on sensitivity in the RA, SA I, and SA II channels than it does in the PC channel.

Effects of the Menstrual Cycle

The sensitivity of a sensory system can be influenced by the menstrual cycle (Parlee, 1983). In vision (Barris, Dawson, & Theiss, 1980) and audition (Henkin, 1974), sensitivity is highest near the time of ovulation. Taste sensitivity is highest during menstruation (Henkin, 1974), and olfactory sensitivity is highest at mid-cycle and during menstruation (Mair, Bouffard, Engen, & Morton, 1978). Cutaneous sensitivity to painful stimuli (Goolkasian, 1980; Robinson & Short, 1977), to changes in skin temperature (Kenshalo, 1966), and to tactile stimulation (Gescheider, Verrillo, McCann, & Aldrich, 1984; Henkin, 1974; Robinson & Short, 1977) have also been found to vary with the menstrual cycle.

An investigation of the effects of the menstrual cycle on the detection of low- and high-frequency vibratory stimuli applied to the thenar eminence (Gescheider et al., 1984) is particularly pertinent to our discussion of the characteristics of tactile channels. In this study detection thresholds for the RA channel (15-Hz stimulus) and the PC channel (250-Hz stimulus) were measured every other day for 44 days. Figure 5.16 shows thresholds plotted as a function of days before and after the start of menstrual flow (M). Thresholds for detecting the 15-Hz stimulus (RA channel) did not change significantly over the menstrual cycle. In contrast, thresholds for detecting the 250-Hz stimulus (PC channel) became progressively lower as menstruation approached. After the onset of menstruation, thresholds of the PC channel gradually increased, reaching their highest point approximately 12–13 days later, near the time of ovulation. Soon thereafter, PC-channel thresholds again began to decrease until the onset of the next menstrual cycle, whereupon they began to increase again. Whether the effects of the menstrual cycle on PC sensitivity are in the periphery at the level of the receptors and their afferent fibers or

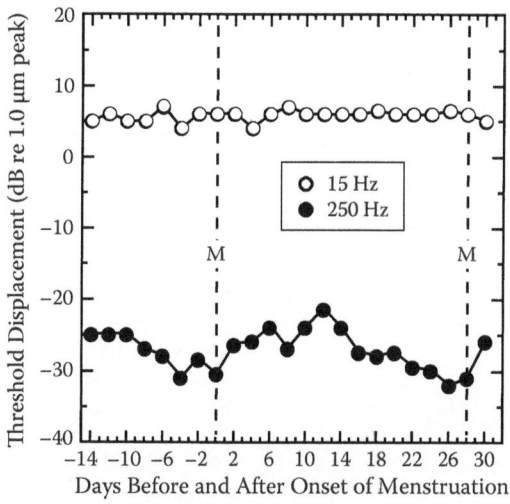

Figure 5.16   Thresholds for the detection of 15-hz and 250-hz stimuli as a function of days before and after the onset of menstruation. Data from Gescheider et al. (1984).

are located within the central nervous system is unknown. Nevertheless, this experiment constitutes yet another demonstration of the unique properties of the PC channel. Although similar to other channels with regard to neural persistence, spatial acuity, temporal acuity, and sensory learning, it differs from the other channels in its unique capability for spatial and temporal summation, its highly tuned frequency selectivity, its greater susceptibility to the deleterious effects of aging, and the effects of the menstrual cycle on its sensitivity.

6
The Functional Roles of Channels

Channels Enhance the Detectability of Stimuli

Before examining the specific role each channel plays in tactile perception, we must first address the general issue of how sensory channels facilitate the overall performance of a sensory system. Acoustic stimuli are processed through separate channels referred to as *critical bands*. Each critical band functions as a filter that passes only the frequencies falling within a relatively narrow range contained within the broader frequency range of an audible sound. Fletcher (1940) suggested that the auditory system contains a bank of such bandpass filters with overlapping frequency ranges. In attempting to detect a signal against a noisy background, an observer will attend to the filter (channel) with a center frequency closest to the frequency of the signal. It is through this channel that the signal is detected. The only noises that can potentially mask the signal being detected are the noise frequency components that pass through this channel. Thus, the filtering property of the channel serves to remove much of the broad-band noise, which, in the absence of the channel, would greatly compromise hearing. The detectability of the signal is limited only by the integrated energy of the frequency components of the noise that pass through the channel. Because of the filtering capacity of the channel, other frequency components of the noise do not contribute to the noise level within the channel, although they can contribute to the noise level within other channels.

A remarkably similar phenomenon has been demonstrated in the tactile sensory system with regard to the properties of its channels. Makous et al. (1995) examined the masking effects of narrow-band vibratory noise applied to the fingertip on the detectability of a sinusoidal signal applied simultaneously to the same site. In this experiment the PC channel was the only channel investigated. As is the case with critical-band filters in the auditory system, Makous and his co-workers found that the PC channel filters and integrates the stimulus energy of the frequency components that fall within the frequency range of the channel. Similar to auditory critical bands, the detection threshold of the signal was determined by the ratio of the energy of the signal to the total energy of the masking noise within the channel, whereas the energy of the noise outside the frequency range of the channel had no effect on the detectability of the signal. The sensitivity of a tactile channel is limited only by the noise occurring within the frequency range of the channel. Thus, the detectability of a stimulus is greatly enhanced by the capacity of a channel to filter noise that would otherwise mask the stimulus and consequently make it more difficult to detect.

Channels Enhance the Discriminability of Stimuli

The importance of independent sensory channels can best be understood by considering the consequences of their absence. Without separate sensory pathways to serve as independent channels, the ability of a sensory system to discriminate among stimuli would be severely limited. This is because the information contained in the sensory pathway of a single channel is highly ambiguous. Rushton's (1972) *principle of univariance* states that a visual stimulus can vary in intensity or wavelength, but the neural response of a visual receptor to that stimulus can vary only in amplitude. If one measures changes in the neural response of a single type of visual receptor exposed to light, it is not possible to know whether the wavelength or the intensity of the light has been changed, because a change in either changes the amplitude of the neural response. The principle of univariance is dramatically illustrated in monochromacy. The retina of a monochromate contains a single type of visual receptor, and as a consequence the individual is unable to detect reliably changes in wavelength, because the same change in receptor potential could be caused either by a change in the wavelength of a light or by a change in its intensity. This ambiguity does not exist in the normal retina, with its three independent color channels (cone types). A visual stimulus of a specific wavelength and intensity will activate to differing degrees each of the three channels. A change in wavelength is encoded as a change in the relative amplitude of the neural response in each of the three channels. A change in intensity, on the other hand, is encoded as a proportional change in the amplitude of the neural response in each of the three channels. Similar coding occurs in olfaction, in which the qualitative differences in smell elicited by odorants are encoded by the relative levels of neural activity in specific types of olfactory receptors (Buck, 1996). Thus, both the visual and olfactory systems encode qualitative changes in a stimulus as a change in the relative levels of activity in different sensory channels. If a sensory system possessed only a single channel, the unambiguous perception of changes in stimulation along variable dimensions would be severely limited, and the number of sensory experiences possible would also be severely limited. Because channels make it possible to detect changes in multiple dimensions of the stimulus, the possible number of sensory experiences in a sensory system with multiple channels becomes vast.

The Functional Roles of the Individual Tactile Channels

It is appropriate at this point to address fundamental questions about the functional role that each channel plays in tactile perception. Accordingly, it is essential to consider the properties of tactile stimuli to which each channel is optimally responsive. An analysis of similarities and differences among channels in this regard should reveal the degree to which channels are specialized in processing specific features of the tactile environment.

All tactile mechanoreceptors respond to mechanical displacement of the skin, and in this respect they are fundamentally the same. But the receptors of the different channels respond differently to skin displacement. For example, as seen in Figure 3.1 and Table 3.1, the nerve fibers innervating the tactile receptors can be classified as rapidly adapting (PC and RA) or slowly adapting (SA I and SA II) in response to mechanical displacement of the skin. Thus, channels can be distinguished by how their receptors and associated nerve fibers respond to mechanical displacement. We shall show that the RA, SA I, and SA II channels can best be described in terms of their responses to changes in the *amplitude* of the mechanical displacement of their receptors, but the PC channel can best be described in terms of its responses to changes in the *energy* of the stimulus. The energy and displacement amplitude of a tactile stimulus are nonlinearly related, and therefore a distinction between a system that responds to energy and one that responds to displacement amplitude must be made. Indeed, this fact has influenced how many investigators specify stimulus intensity in psychophysical and neurophysiological experiments. Although it is not difficult to convert units of stimulus amplitude into units of stimulus energy, the theoretical implications of choosing one over the other as a measure of stimulus intensity critically define how the channels are conceptualized.

The Decibel Scale

The intensity of a vibratory stimulus is generally measured as the amplitude of displacement of the skin resulting from the application of a vibrating object to it. Thresholds measured in displacement amplitude, however, are often converted to energy by squaring them. The squared displacement amplitude, which is proportional to energy, can be expressed in decibels (dB) by the formula

$$dB = 10 \log (A^2/A^2_{ref})$$

where A is displacement amplitude and A_{ref} is an arbitrary reference amplitude, such as the value of 1 micrometer used in many vibrotactile experiments.

Thus, the decibel scale consists of a logarithmic scale of the ratio of two energies. There are two major reasons why investigators express thresholds in decibels—one is simply convention and the other is theoretical, with the underlying assumption being that it is energy and not amplitude to which a tactile system responds. But what is the evidence that the tactile system responds to stimulus energy? The answer to this question is not simple, but the evidence strongly suggests that the PC channel can best be conceptualized as an energy-detecting system, whereas the RA, SA I, and SA II channels are best thought of as detecting displacement amplitude rather than energy. This fundamental difference between the PC channel and the other three channels helps reveal the functional role that each channel plays in the perception of tactile stimuli.

The Functional Role of the PC Channel

The hypothesis that the PC channel behaves as an energy-integrating system is supported by several diverse experimental findings. In studies of spatial summation in the PC channel, detection thresholds expressed in energy terms (amplitude squared) have been found to decrease in proportion to the size of the skin area stimulated (e.g., Gescheider, 1976; Gescheider et al., 2002; Verrillo, 1963). It can be seen in Figure 6.1A, for example, that as the size of the contactor delivering the stimulus to the thenar eminence increases, there is a proportional decrease in threshold expressed as amplitude squared (a value directly proportional to energy). The solid curve represents the expected decrease in amplitude squared if doubling the size of the contactor halves the energy required for stimulus detection. The fit of the data points to this function is remarkably good. In contrast, although the same threshold measurements expressed in displacement amplitude rather than amplitude squared clearly decrease with increases in the size of the contactor, they fail to match the function expected if doubling the size of the contactor halves the displacement amplitude required for stimulus detection (Figure 6.1B). This shows that the PC channel responds primarily to stimulus energy rather than to stimulus amplitude and integrates stimulus energy over the area of stimulation at detection threshold. The reciprocal relationship between stimulus energy at threshold and the size of the skin area stimulated suggests that total spatial integration of stimulus energy occurs in the PC channel. Accordingly, area and energy (amplitude squared) can be traded one for the other (i.e., doubling stimulus area while halving stimulus energy) at detection threshold in the PC channel. The relationship

$$A^2 \times S = \text{Constant}$$

expresses this principle by stating that, at detection threshold, the energy level of the stimulus (A^2) times the size (S) of the stimulus remains constant, regardless of the size of the contactor. Thus, the *total energy* of the stimulus inte-

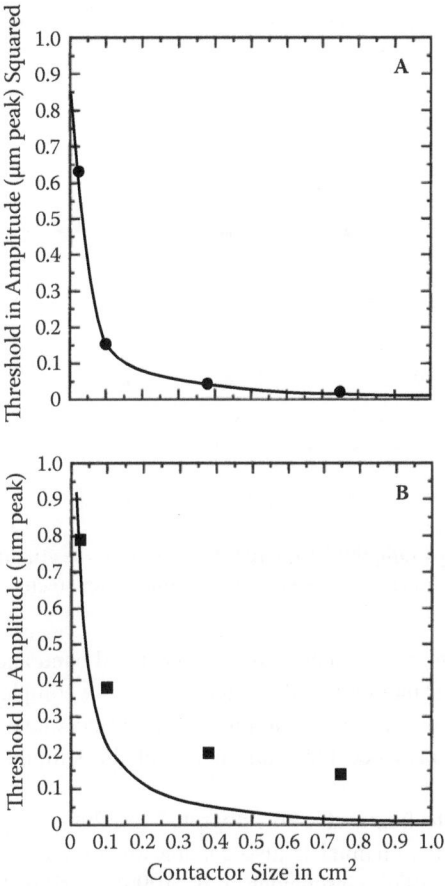

Figure 6.1 Detection thresholds expressed in terms of energy (amplitude squared) (A) or in terms of amplitude (B) plotted as a function of the size of the stimulus (contactor size) applied to the thenar eminence. The solid curves represent expected results for total energy integration over area (A) and total amplitude integration over area (B). Data from Gescheider et al. (2002).

grated over its size ($A^2 \times S$) remains constant at detection threshold as the size of the stimulus is varied (see Figure 6.2).

Total spatial integration of stimulus energy also occurs in the visual system, and the reciprocal relation at detection threshold between the energy level of the stimulus and its size is known as *Ricco's law*. The maximum area over which complete energy integration occurs in vision is less than one degree of visual angle in the peripheral retina, and even less in the fovea (Graham, Brown, & Mote, 1939). Similarly, in hearing, complete energy integration occurs only within the spatial areas on the basilar membrane comprising criti-

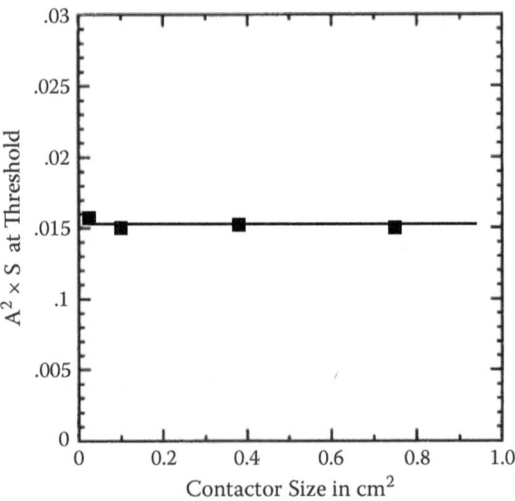

Figure 6.2   Energy (amplitude squared) times size of the stimulus ($A^2 \times S$) at detection threshold as a function of the size of the stimulus (contactor size).

cal bands. Within the frequency range of a critical band and its corresponding area on the basilar membrane, the detectability of a complex stimulus consisting of several pure tones presented together is independent of the number of tones presented, provided the total energy of the stimulus remains constant (Gässler, 1954).

In addition to being able to integrate stimulus energy over the area of stimulation, the PC channel is also capable of integrating energy over time (e.g., Gescheider, 1976; Gescheider et al., 1999; Verrillo, 1965). In studies of temporal summation in the PC channel, detection thresholds, when expressed in energy terms (amplitude squared), have been found to decrease in proportion to the duration of the stimulus. Total energy integration occurs up to approximately 80 ms, after which partial integration is observed up to about one second. It is seen in Figure 6.3A that as the duration of a stimulus applied to the thenar eminence increases, the threshold expressed in energy terms (amplitude squared) decreases proportionally, thus showing that total energy integration occurs over time. The solid curve represents the function expected if doubling the duration of the stimulus halves the energy required for stimulus detection. This function fits the data points quite well. In contrast, when the same threshold measurements are expressed in displacement amplitude, they clearly decrease, but do not fit the function expected if doubling stimulus duration halves the displacement amplitude required for stimulus detection (solid curve in Figure 6.3B). As seen in Figure 6.4, there is a reciprocal relationship between stimulus energy and the duration of the stimulus at threshold, such

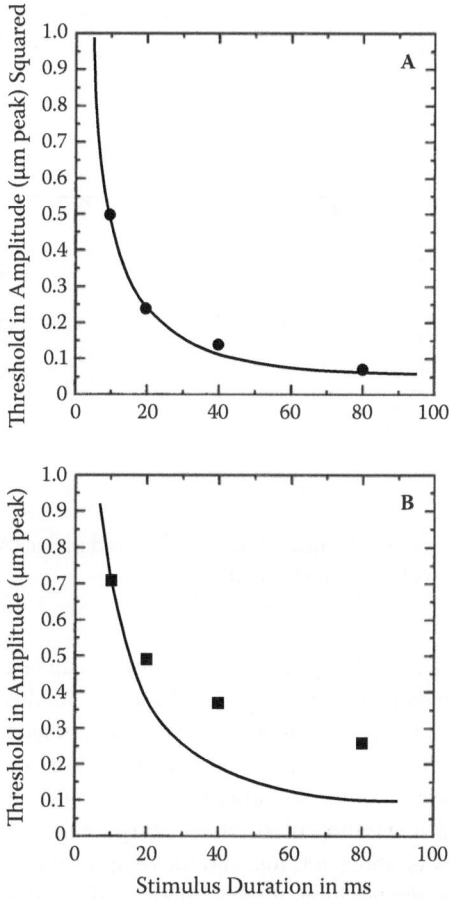

Figure 6.3 Detection thresholds expressed in terms of energy (amplitude squared) (A) or in terms of amplitude (B) plotted as a function of the duration of the stimulus applied to the thenar eminence. The solid curves represent expected results for total energy integration over time (A) and total amplitude integration over time (B). Data from Gescheider et al. (2002).

that the product of the stimulus energy level at threshold (A^2) times the duration of the stimulus (D) remains constant for the detection of stimuli up to about 80 ms in duration. The relationship written as

$$A^2 \times D = \text{Constant}$$

for temporal summation in the PC channel is analogous to the equation $A^2 \times S =$ Constant, obtained for spatial summation in the PC channel. Within the PC

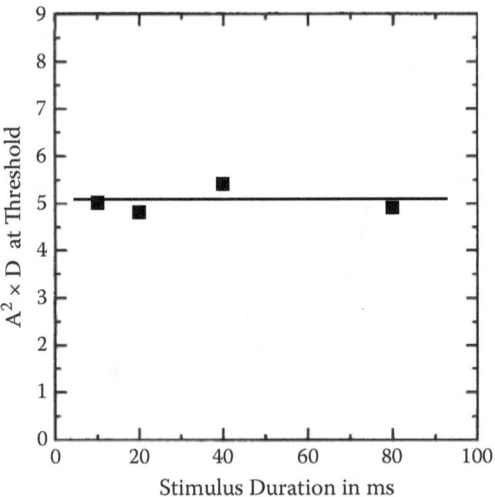

Figure 6.4   Energy (amplitude squared) times stimulus duration ($A^2 \times D$) at detection threshold as a function of stimulus duration.

channel the duration and energy level of the stimulus can be traded one for the other (i.e., doubling stimulus duration while halving stimulus energy level) for the detection of stimuli up to about 80 ms in duration; thus the *total energy* of the stimulus integrated over its duration ($A^2 \times D$) remains constant at detection threshold as the duration of the stimulus is increased up to about 80 ms.

Total temporal integration of stimulus energy also occurs in the visual system up to a limit of about 100 ms, and the reciprocal relationship between the energy level and the duration of the stimulus at detection threshold known as *Bloch's law* has been repeatedly verified (Barlow, 1958; Graham & Margaria, 1935; Karn, 1936). In hearing, near total energy integration over stimulus duration has been found to occur up to about 100–200 ms (Garner & Miller, 1947; Jeffress, 1975).

Additional support for the hypothesis that the PC channel behaves as an energy integrator comes from the results of a study of tactile masking in which observers detected a 200-Hz sinusoidal stimulus in the presence of vibratory noise of varied intensity and bandwidth (Makous et al., 1995). It was found that the amount by which the detection threshold was elevated by a masking noise within the PC channel was determined by the total integrated energy of the frequencies comprising the noise. Thus, as in spatial and temporal summation, the PC channel can best be understood as an energy-integrating system, but in this study the PC channel integrated the energy from the frequency components of a noise masker.

The energy-detection model of the PC channel is further supported by the results of masking experiments in which both sinusoids and noise were used as masking stimuli (Gescheider et al., 1989; Hamer et al., 1983). In these experiments the sinusoidal and noise maskers raised the threshold for detecting a 250-Hz test stimulus by the same amount. This occurred, however, only when the elevation in the detection threshold produced by the masker (noise or sinusoid) was expressed as an increment in energy resulting from the addition of the energy of the test stimulus to the energy of the masking stimulus. The detection of energy increments appears to be the underlying mechanism responsible for stimulus detection in the PC channel. A similar effect is observed in hearing if the elevation in the threshold measured in the presence of a masking stimulus is expressed in terms of energy increments (Raab, Osman, & Rich, 1963). In this study the test stimulus and the masking stimulus were both noise. When the test and masking stimuli were correlated noise with identical waveforms and in phase, less masking occurred than when they were uncorrelated with independent and different waveforms. But when the test stimulus was defined as an increment in energy added to the energy of the masking noise, masked thresholds for the correlated and uncorrelated noise stimuli were the same. These results support the notion that in both audition and the PC channel in touch, the mechanism underlying stimulus detection is the sensing of increments in stimulus energy.

The model of the PC channel as an energy-integrating system applies to suprathreshold levels of stimulation as well. This conclusion is supported by the results of an experiment in which suprathreshold sinusoidal vibrations of different frequencies were combined to produce a complex wave (Marks, 1979). The frequencies selected were 200 Hz and 250 Hz, both of which selectively excite the PC channel at the intensity levels used in the experiment. Each of seven intensity levels at 200 Hz was combined factorially with each of seven intensity levels at 250 Hz to produce 49 different complex stimuli. The observer's task was to estimate the subjective magnitude of the intensity of each of the 49 complex stimuli. Plotted in Figure 6.5 is the relationship between estimates of sensation magnitude and the total energy of each of the 49 complex stimuli. Although there are different ways to specify the physical intensity of the 49 complex stimuli used in this experiment, only when intensity was specified in terms of total stimulus energy was it possible to generate a function that fit precisely the subjective magnitude estimates made by the observers. From these results, Marks concluded that the judged sensation magnitude of a complex stimulus whose frequency components selectively excite the PC channel is determined by the sum of the energies of the individual components weighted by their effectiveness in stimulating this channel.

The psychophysical experiments on spatial summation, temporal summation, masking, and suprathreshold sensation magnitude described above show

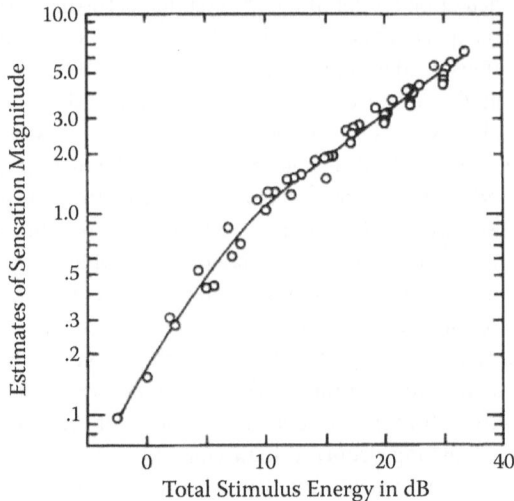

Figure 6.5 Estimates of sensation magnitude as a function of the total weighted energy of a two-component stimulus. Zero decibels represents the relative energy of a single 250-Hz stimulus at an intensity level of -16 dB re 1 μm. From Marks (1979).

that the PC channel acts as an energy integrator. However, it should be kept in mind that the receptors of this channel, Pacinian corpuscles, as well as all other mechanoreceptors, respond directly to mechanical displacement of the skin and not directly to stimulus energy. Thus, at the input stage of the PC channel, mechanical displacement of Pacinian corpuscles is converted to depolarizing receptor potentials, which in turn trigger action potentials in peripheral afferent fibers that are conducted to the central nervous system for further sensory analysis. At some stage in this process, the PC channel begins to behave as an energy integrator. Although at this time the exact location at which this occurs is unknown, we do know that the PC channel's ability to integrate energy is what makes it ideally suited to detect minute mechanical disturbances of the skin. Indeed, the ability to summate spatially and temporally enhances the sensitivity of this channel substantially. The fact that spatial and temporal summation are exclusive properties of the PC channel is one of the reasons why the sensitivity of this channel is unmatched by any of the other three. Hence, it appears that the functional role of the PC channel is to detect vibration, particularly high-frequency vibration, with remarkable sensitivity, and this is achieved through spatial and temporal integration of stimulus energy. We turn now to the functional roles of the RA, SA I, and SA II channels.

The Functional Role of the RA Channel

It has been argued that the RA channel is superbly suited to detect changes in the waveform of vibratory stimulation applied to the skin (Hollins, Delemos, & Goble, 1996). Vibratory waveforms are described in terms of changes in amplitude of displacement over time. Hollins and his co-workers note that frequency discrimination—a form of waveform discrimination—is relatively good at low vibration frequencies, where the RA channel is strongly activated, and relatively poor at high frequencies, where the PC channel is strongly activated. They also point out that the relatively flat frequency-selectivity function of the RA channel makes it much more suitable for waveform discrimination than the PC channel with its highly tuned frequency-selectivity function. To encode effectively the waveform of a complex stimulus made up of many frequency components, a channel must be able to respond equally well to all frequency components of the stimulus. If the frequency-selectivity function of a channel is highly tuned, the encoding of a complex stimulus waveform will be distorted to the extent that the frequency components of the stimulus are differentially filtered.

The hypothesis that the RA channel processes information about the amplitude rather than the energy of a vibratory stimulus is supported by the finding that the amount of adaptation of the RA channel in detecting a 20-Hz stimulus is determined by the peak amplitude of the adapting stimulus rather than by its energy. Specifically, a complex stimulus consisting of the combination of two sinusoidal vibrations with frequencies of 90 Hz and 110 Hz was found to have the same adapting effectiveness as a 20-Hz or a 100-Hz adapting stimulus of the same peak amplitude, even though the energy level of the compound adapting stimulus was much higher than that of the 20-Hz or the 100-Hz adapting stimulus. This finding demonstrates that the neural mechanism of the RA channel responsible for adaptation encodes the peak amplitude rather than the energy of the stimulus. According to Hollins et al. (1996), the RA channel, by being keenly attuned to amplitude, is capable of precise frequency and waveform discrimination. Thus, the RA channel provides information about the temporal structure of a vibratory stimulus that is characterized by its amplitude, frequency, and complexity rather than by its energy. According to Hollins and his co-workers (1996), "Analysis of stimulus dynamics, rather than mere detection, appears to be a primary function of this channel."

The Functional Role of the SA I Channel

The absence of spatial and temporal summation in the SA I channel suggests that this channel also processes information about the amplitude rather than the energy of mechanical deformations of the skin. The SA I channel appears to contribute to the processing of spatially distributed variations in the amplitude of a tactile stimulus. This property of the SA I channel may help explain why human observers are able to perform so well in tactile spatial-acuity tasks.

There are several factors that could influence the spatial acuity of a channel, such as the density of its receptors, the size of the receptive fields of its afferent nerve fibers, the variability in the sensitivity of the afferent fibers, and the noise in the system. Given this complexity, the question of which afferent nerve fibers are best at encoding spatial information—and how they do it—is not easy to answer. Although this problem has been the focus of much research, the possibility must not be overlooked that more than one fiber type contributes to tactile spatial processing. For example, Chapter 7, on the interaction of channels, presents evidence that the perception of tactile roughness can in some circumstances involve contributions from both the SA I and PC channels.

The most promising way to determine the role each channel plays in spatial perception is to study how a *population* of neurons within a channel responds to charges in the spatial parameters of a tactile stimulus. In general, this approach consists of determining: (1) the neurophysiological responses of single afferent nerve fibers to changes in the spatial properties of a tactile stimulus, such as its location, shape, and motion; (2) the changes in the population response profile of many neurons as spatial aspects of the stimulus are changed; and (3) the correlation between changes in the population response profile and the performance of observers in a spatial-discrimination task.

This approach was utilized by Wheat, Goodwin, and Browning (1995) in an attempt to discover the peripheral neural mechanisms underlying the capacity to determine the positions of objects applied to the skin. These investigators measured the ability of observers to discriminate the positions of spherical surfaces applied to the finger pad. The procedure consisted of applying a curved stimulus to a position on the finger pad, followed by presentation of the same stimulus to either the same position or a different position. The observer's task was to report whether the second stimulus presentation was to the same or to a different location on the skin. The discrimination thresholds for responding correctly 75 percent of the time averaged 0.55 mm between skin locations for a moderately curved sphere and 0.38 mm between skin locations for a sphere of greater curvature. Because these values are substantially lower than those predicted from an SA I nerve-fiber density of about one fiber per square millimeter, they constitute examples of hyperacuity (Loomis, 1979; Loomis & Collins, 1978) and cannot be accounted for by the characteristics of single receptors. The receptive field of a single fiber is simply too large for the fiber to encode unambiguously a change in the position of the stimulus. Hence, the position of the sphere must be encoded in the responses of a population of nerve fibers activated by the curved stimulus. To test this hypothesis, Goodwin et al. (1995) recorded from single afferent nerve fibers in the monkey to obtain measures of the neural responses of the fibers in a population of afferents with receptive-field centers of varying distances from the stimulus. The same spherically shaped stimuli that had been used to measure psychophysically the performance of human observers were used to measure the receptive-field profiles of SA I, RA, and PC afferent fibers in the monkey. The number of action potentials occurring in the first second of stimulus application was recorded.

Shown in Figure 6.6 are three-dimensional profiles of the average normalized neural response as a function of the position of the stimulus within an SA I receptive field. The Y distance is the proximal-distal distance along the finger pad, and the X distance represents positions across the fingertip. The zero points represent the center of the receptive field. It is apparent that the receptive-field profile of an SA I fiber becomes higher and narrower as the amount of curvature in the stimulus applied to the receptive field increases. This neurophysiological finding corresponds well to the psychophysical finding that the localization of a stimulus on the fingertip is more precise when the stimulus has highly curved surfaces (Wheat et al., 1995). Although the correlation between the psychophysical and neurophysiological data from single SA I fibers is striking, the problem remains to explain how an observer can discriminate a change in the location of a stimulus from a change in the force with which it is applied. This cannot be accomplished by a single SA I fiber because both a change in the location and a change in the intensity of the stimulation cause a change in the magnitude of neural activity in a single SA I fiber. Thus, signals from a single SA I fiber are ambiguous as to their source. Only through population coding can this ambiguity be resolved. For example, proportional increases in the neural responses of SA I fibers A and B with overlapping receptive fields can encode an increase in stimulus intensity. With these same fibers, an increase in the neural response of fiber A and a concomitant decrease in the neural response of fiber B can encode a change in the location of the stimulus toward the center of the receptive field of fiber A and away from the center of the receptive field of fiber B. Of course, this is an oversimplified example involving a population of only two SA I fibers. Given that the density of SA I fibers on the pad of the fingertip is approximately one fiber per square millimeter, and the width of their receptive fields averages approximately 2–4 mm, the overlap in receptive fields is substantial, and consequently the number of fibers activated by the spherical stimuli used by Goodwin et al. (1995) was quite large. As pointed out by Goodwin and his co-workers, "If all the SA Is had identical properties, then the population profile would be identical to the single afferent profiles." Under these conditions, as demonstrated by their recordings from single SA I afferent fibers, the shape of the population profile would change as the curvature of the stimulus is changed, whereas the height of the profile would change as the force with which the stimulus is applied is changed. Such changes in the population profile account for the fact that human observers can discriminate

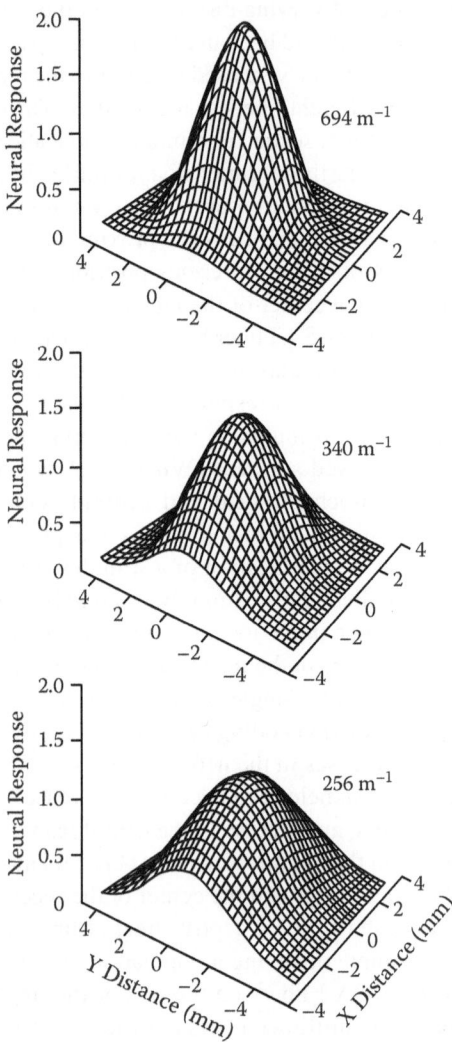

Figure 6.6   Three-dimensional neural-response profiles for SA I fibers in the monkey. The profiles show the normalized average neural response as a function of the position on the finger pad of stimuli of three different curvatures ranging from 694 m^{-1} (highest curvature) to 256 m^{-1} (lowest curvature). From Goodwin et al. (1995).

between a change in the shape and a change in the contact force of a stimulus applied to the skin (Goodwin & Wheat, 1992).

Goodwin et al. (1995) report that RA fibers also responded to the spherical stimuli used in their study, but the responses were weaker than observed in SA I fibers. They concluded that although RA fibers may not provide much information that can be used to determine shape or contact force during passive touch, they may, because of their small receptive fields, high innervation density, and sensitivity to movement, provide such information in tasks in which the finger moves over an object or the object is moved over the skin, as in the work of LaMotte and Srinivasan (1987). Thus, the RA channel appears to contribute to the perception of the shapes of objects moving over the surface of the skin.

The argument that the SA I channel is superbly able to process the spatial details of a tactile stimulus is supported not only by the work described above on stimulus localization but also by the results of experiments on the detection of the edges and gaps that constitute the distinctive features of objects (Wheat & Goodwin, 2000). It has been found that the observer's ability to detect changes in the width of a groove in a block applied to the fingertip can be accounted for by the population-response profile of SA I nerve fibers. The difference limen for discriminating a change in groove width was approximately 0.2 mm—a value sufficiently small to constitute a case of hyperacuity requiring explanation in terms of the population response of a significant number of SA I fibers with overlapping receptive fields. The population response of SA I afferent fibers was simulated from recordings of the responses of single SA I fibers as the grooved block was presented at different positions within the fiber's receptive field. The simulated SA I population response to the block with a groove width of 2.8 mm is seen in Figure 6.7. The groove was presented

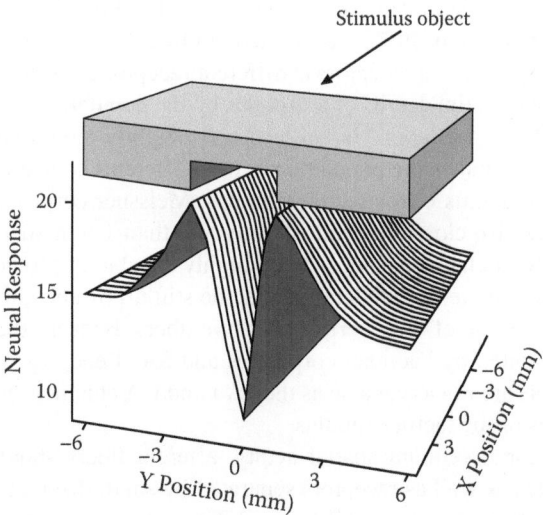

Figure 6.7   Simulated SA I population response to a stimulus with a groove width of 2.8 mm. From Wheat and Goodwin (2000).

perpendicular to the long axis of the finger. The nerve fibers in the simulated population that innervate locations on the skin along the fingertip differ dramatically in their response to presentation of the grooved stimulus. The response is lowest for neurons located in the center of the stimulus groove, highest for neurons at the edges of the groove, and moderate for neurons on either side of the edges. This neural population response is consistent with the psychophysical results obtained with groove widths that are easily detected. Specifically, it appears that the spatial elements of a grooved surface are encoded by the neural responses, including enhanced response to the groove edges, by a population of afferent fibers in the SA I channel.

It should be pointed out that when comparing psychophysical performance with the neural responses from a simulated population of nerve fibers, it is prudent to allow for uncertainty in such factors as innervation density, variability in sensitivity across fibers, and neural noise (Goodwin & Wheat, 2004). Nevertheless, the results of this simulation of the SA I population response to edges and gaps and the results obtained for stimulus localization leave little doubt that the peripheral afferent fibers of the SA I channel are capable of remarkable performance in processing information about the spatial characteristics of tactile stimuli.

This conclusion is entirely consistent with the principle that receptor density, which is relatively high for SA I nerve fibers, correlates highly with tactile spatial acuity. For example, the progressive increase in receptor density from palm to fingertip that exists in glabrous skin (Johansson & Vallbo, 1979) correlates highly with the corresponding increase in the ability of the observer to resolve the spatial details of stimuli applied to the skin (Craig & Lyle, 2001, 2002; Stevens & Choo, 1996). As pointed out by Johnson and Phillips (1981), it is the spacing between receptors at different receptor densities that is critical in encoding the spatial details of a stimulus by the population of afferent nerve fibers. The close spacing of Merkel receptors (slightly more than 1 mm apart on the fingertip) makes the population of SA I afferents well suited for encoding the spatial details of the tactile stimulus. Meissner corpuscles, with their RA fibers, are also closely spaced (slightly less than 1 mm apart). Thus, both the RA and SA I channels appear to be highly capable of providing information about the minute spatial details of tactile stimuli by encoding them in the population response of their peripheral nerve fibers. Because the skin is more sparsely populated by Pacinian corpuscles and SA II end organs, the PC and SA II channels are not as capable as the SA I and RA channels of encoding the spatial details of the tactile stimulus.

Ideally, for maximum spatial acuity, afferent fibers should have small receptive fields as well as receptors separated by small distances. In this way, individual features such as Braille raised dots, which are separated by small distances, can independently excite separate nerve fibers, thus encoding the dot pattern in the population response of the afferent nerve fibers. This is essentially what Connor et al. (1990) demonstrated by recording the neural responses of SA I, RA, and PC nerve fibers as raised dot patterns were presented to the monkey's fingertip. It was found that because of their small receptive fields, SA I and RA fibers were able to differentiate between raised dots and the blank space between them when the spacing of the dots exceeded the receptive-field diameter of the fibers. When the dots were brought closer together, the fibers were unable to resolve the individual dots, thus demonstrating that fibers cannot discriminate between a single dot and two dots when both are located within the same receptive field. Because of their large receptive fields, individual PC fibers were relatively poor at resolving the dots in the pattern.

Receptive-field size and distance between receptors may set limits on spatial resolution, but the inferior performance of PC fibers in no way rules out their participation in the spatial perception of stimuli that clearly exceed the psychophysical limits of resolution. Recall that when channels are psychophysically isolated by carefully choosing the appropriate stimulus intensity, frequency, and contactor size, spatial discrimination in the PC channel is only slightly inferior to that in the RA and SA I channels (Sherrick et al., 1990) or even equal to it (Rogers, 1970).

The Functional Role of the SA II Channel

Less is known about the functional role of the SA II channel than about the other three channels, although it is well established from psychophysics and recordings from the afferent nerve fibers of humans that the channel exists. The absence of information about the SA II channel is due in part to the fact that monkeys, from whom so much neurophysiological information has been collected, do not possess SA II nerve fibers (Darian-Smith & Kenins, 1980; Darian-Smith & Oke, 1980; Lamb, 1983; Lindblom, 1965), and this has severely limited the amount of relevant neurophysiological research on this channel. Unfortunately, there are few studies in which electrophysiological recordings have been taken from human SA II fibers. Further, as discussed in Chapter 3, the mechanoreceptive end organs of the SA II channel in glabrous skin have not been conclusively identified. Large numbers of SA II afferents are found in both hairy and glabrous skin, but Ruffini endings, proposed by many investigators to be the SA II end organ (e.g., Johansson & Vallbo, 1979), are found in hairy but apparently not in glabrous skin (e.g., Dellon, 1981; Paré et al., 2003; Paré et al., 2002). Further anatomical research is clearly needed to resolve this issue. If Ruffini endings function as SA II end organs in hairy skin but are confirmed to be absent in glabrous skin, then some other mechanoreceptor must function as the SA II end organ in glabrous skin. This raises the intriguing question of whether the SA II channel is different from the other three channels by having two different end organs—the Ruffini ending in hairy skin and a different and currently unidentified end organ in glabrous skin. If this turns out to be the case, then differences in the psychophysical performance of the SA II channel in hairy and glabrous skin could well result from the fact that the SA II end organs are different, and the functional role of the SA II channel in hairy and glabrous skin may be different as well.

Nevertheless, it can be stated, based on the small amount of work that has been conducted, that it is unlikely that the SA II channel contributes significantly to texture perception. Phillips, Johansson, and Johnson (1992) have reported that all tactile mechanoreceptive afferents of humans except SA II afferents respond vigorously to textured surfaces. It has been suggested that the primary function of SA II afferents is to provide information about the detection, magnitude, and rate of change of tensions within the skin and between the skin and deeper, less flexible structures (Johansson, 1978). The activation of SA II fibers resulting from skin stretch induced by moving an object across the skin may contribute to the tactile perception of movement (Essick, 1998). The SA II channel may also contribute to the perception of hand conformation by providing a neural image of skin stretch over the entire hand (Johnson et al., 2000).

Specialization of Channels

Our descriptions of the properties of tactile stimuli to which each channel is optimally responsive have shown that channels are specialized for processing specific types of information. The PC channel attains its exceptional capacity for detecting high-frequency vibration through the process of energy integration. In contrast, the RA channel, rather than responding to stimulus energy, detects the waveform of the vibratory stimulus by responding to changes in its displacement amplitude over time. The SA I channel detects spatial details by responding with remarkable acuity to spatially distributed amplitudes of skin displacement. Finally, the SA II channel appears to be highly responsive to the stretching of the skin.

The high degree of specialization that channels exhibit in responding to tactile stimulation could lead one to conclude that channels operate as *feature detectors* that respond vigorously to the presence of some features of the stimulus, while being completely unresponsive to others. According to a feature-detection model, the perception of a complex tactile stimulus results from the sum of all of the individual features encoded by independent and highly specialized channels. The problem with this model of tactile perception is that features of the tactile stimulus, although detected best by a particular channel, can also be detected by the other channels, although with less precision. For example, as described in Chapter 5, localization of a stimulus is almost as accurate in the PC channel as it is in the RA and SA I channels (Sherrick et al., 1990). The alternative—and in our view correct—model is one in which more than one channel, and in some cases all channels, are capable of contributing information about a particular stimulus feature. Accordingly, the tactile percept results not from the sum of features processed by different channels but instead from the blending of activity from all channels. This channel-interaction hypothesis is the topic of Chapter 7.

7
Channel Interactions

Introduction

The results of our analysis of psychophysical and neurophysiological data support the hypothesis that four separate information-processing channels, each with its own specific receptors and peripheral nerve fibers, are responsible for the perception of tactile stimuli. However, fundamental questions remain concerning the nature of how these channels, with their individual properties, operate together in the perception of tactile stimuli encountered in the natural world. We hypothesize that suprathreshold stimulation at an intensity level high enough to activate all of the channels produces tactile perceptions that represent integration within the central nervous system of the activities of the separate channels. Evidence supporting this hypothesis will now be considered.

Summation of Sensation Magnitude Across Channels

The initial support for the hypothesis that channels interact at suprathreshold levels of stimulation comes from the finding that the perceived intensity of two brief stimuli presented simultaneously or in rapid succession is equal to the sum of the perceived intensities of each stimulus presented alone, but only if they activate separate channels (Marks, 1979; Verrillo & Gescheider, 1975). In contrast, when the two stimuli activate the same channel, the perceived intensity of the pair is determined not by the sum of their individual perceived intensities but by the total energy of the two stimuli (Marks, 1979; Verrillo & Gescheider, 1975). In one experiment, the sensation magnitude of a pair of 20-ms stimuli was measured by requiring the observer to adjust the sensation magnitude of a matching stimulus, presented after the pair, to the total sensation magnitude of the pair (Verrillo & Gescheider, 1975). In Figure 7.1 the difference in decibels

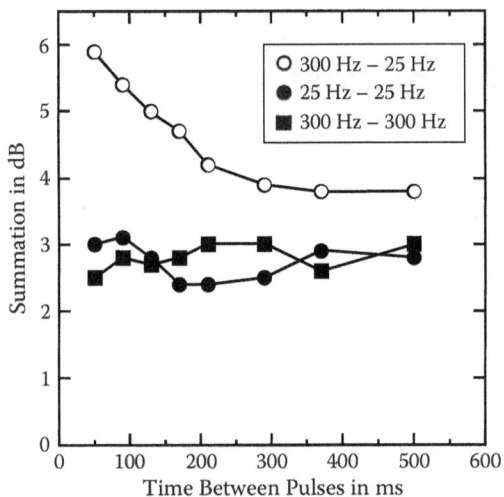

Figure 7.1   Summation of sensation magnitude across the PC and RA channels (300 Hz-25 Hz), within the RA channel (25 Hz-25 Hz), and within the PC channel (300 Hz-300 Hz). From Verrillo and Gescheider (1975).

between matches to the pair and matches to the second member of the pair is plotted as a function of the time interval between the first and second stimulus of the pair. When the time interval between stimuli in the pair was short and the stimuli were in separate channels (300 Hz for the PC channel and 25 Hz for the RA channel), the observer set the matching stimulus 6 dB higher when matching to the sensation magnitude of the pair than when matching to the second stimulus in the pair. This 6-dB difference corresponds exactly to a doubling of sensation magnitude as measured by magnitude estimation (Verrillo et al., 1969). Figure 7.1 also shows that this effect is seen only when the time interval between pulses is very short. When the two stimuli in the pair were 25 Hz (both within the RA channel) or when they were both 300 Hz (both within the PC channel), the observer adjusted the matching stimulus to be 3 dB higher when matching to the pair than when matching to the second stimulus in the pair. This 3-dB difference represents a doubling of stimulus energy.

These findings indicate that whereas perceptual interactions within a channel occur at the level of the stimulus in the form of the summation of their energies, interactions between channels occur at a later stage of processing, after stimulus energies have been transformed into perceptual magnitudes. The similarity between sensation-magnitude summation across tactile channels and loudness summation across auditory critical bands is striking. According to critical-band theory (Scharf, 1970; Zwicker, Flottorp, & Stevens, 1957; Zwicker & Scharf, 1965), the overall loudness of a sound can be calculated by summating loudness over all of the critical bands excited by the acoustic stimulus. Specifically, within each critical band, the total energy of the frequency components of the sound exciting the band is transformed into perceived loudness, and total loudness results from summation across bands of the loudness in each. Thus, for tactile channels and auditory critical bands, the perceived sensation magnitude of a stimulus is determined by linear summation across channels of the sensation magnitudes generated within each channel. Furthermore, in both touch and hearing, the sensation magnitudes generated within channels are determined by the total energy of the frequency components of the stimuli activating the channel.

The logic of determining whether summation occurs at a stage after the stimulus has been transformed by the nervous system into sensation magnitude or whether it occurs prior to any such transformation is shown in Figures 7.2 and 7.3, which illustrate the two possible ways in which stimulus energy can be converted into sensation magnitude. This transformation is made separately for each component of a complex stimulus, provided the components are in separate channels. According to a linear-summation model, total sensation magnitude is the sum of the sensation magnitudes of the components of the complex stimulus (Figure 7.2). In contrast, when two stimulus components are presented within the same channel, it is their energies that summate, and the predicted increase in sensation magnitude in this instance (Figure 7.3) is much

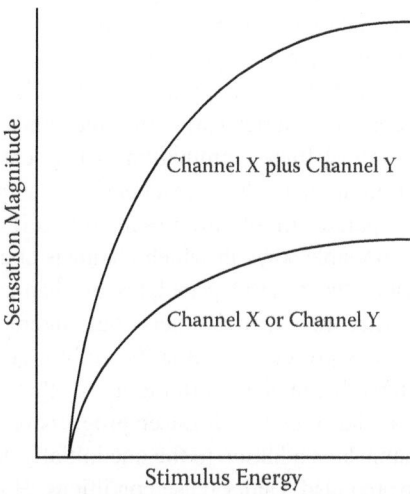

Figure 7.2 Sensation magnitude as a function of stimulus intensity for channel X or channel Y activated separately. Summated sensation magnitude as a function of stimulus intensity when channel X and channel Y are activated simultaneously.

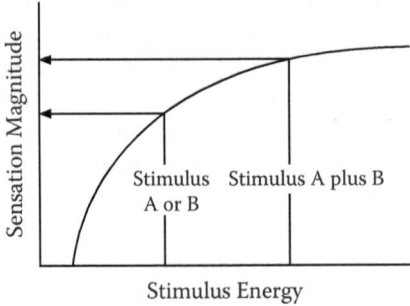

Figure 7.3   Summation of stimulus energy within a channel when stimulus A and stimulus B are presented simultaneously.

less than a doubling of sensation magnitude. In this model the components of a complex stimulus add their energies within a channel and add their sensation magnitudes across channels (Verrillo & Gescheider, 1975).

The results of a direct test of the channel-summation model are seen in Figure 7.4 (Marks, 1979). In support of this model is the finding that at all stimulus-intensity levels, the estimate of the sensation magnitude of a complex stimulus with PC (250-Hz) and non-PC (20-Hz) component stimuli is essentially equal to the sum of the magnitude estimates of the components presented separately. The fine dashed-line function represents the predicted values for linear summation between the two channels.

A test of the hypothesis that the perception of any complex tactile stimulus involves the interaction of channels requires that we fully understand the characteristics of each channel and determine the rules under which the channels interact. The discovery of linear summation of the sensation magnitudes of stimuli exciting different channels represents one example of how the channels interact when separate stimuli are presented to different channels. There are, however, other possible ways in which channels could interact that must be investigated before the general principles of channel interaction can be stated. One of the other ways that channels could interact is when a stimulus of a single frequency is presented at a sufficiently high intensity to activate more than one channel. It can be seen in Figure 4.3 that as the intensity of the stimulus at any particular frequency is raised progressively above the detection threshold, other channels in addition to the one initially detecting the stimulus are eventually activated also. Under these conditions, Hollins and Roy (1996) found that channels interact by partially rather than completely summating their sensation magnitudes. Specifically, the overall sensation magnitude of a vibratory stimulus of a single frequency applied to the fingertip was found to be equal to the sensation magnitude in either the RA or PC channel, whichever

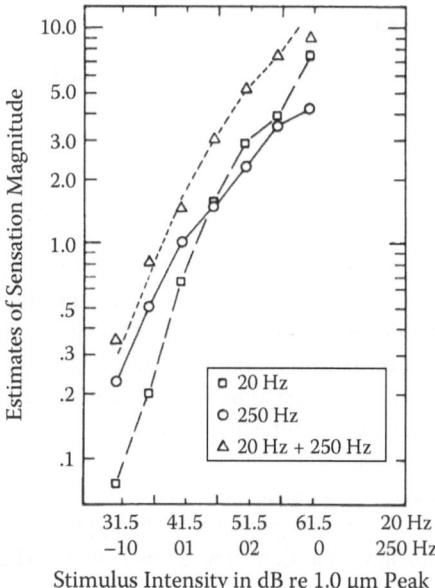

Figure 7.4 Magnitude estimation of the sensation magnitude of 20-Hz and 250-Hz stimuli presented separately, and magnitude estimates of their sensation magnitude when presented together. From Marks (1979).

was greater at a particular frequency, plus approximately half the sensation magnitude in the more weakly activated channel. Thus, the rules by which channels do or do not interact seem to be highly dependent on the composition of the stimulus.

But why does complete summation occur when two separate stimuli at widely differing frequencies are presented either sequentially (Verrillo & Gescheider, 1975) or simultaneously (Marks, 1979), whereas only partial summation occurs when one stimulus of a single frequency is presented? According to Hollins and Roy (1996), the answer may depend on whether the somatosensory system detects two distinct stimuli or only a single stimulus. In this regard, it is our hypothesis that channels interact more effectively when the observer can easily attend to each—a process that should be greatly facilitated when the channels are activated by separate, highly discriminable stimuli, as in Marks (1979) and Verrillo and Gescheider (1975). As will soon become apparent, this hypothesis gains further support from the results of experiments on texture perception, in which it was found that the interaction of the SA I and PC channels in the perception of roughness depends on whether the observer can attend simultaneously to both of these channels when feeling a textured surface.

The Perception of Texture

The generality of the principle that channels interact in the perception of suprathreshold stimuli has been extended recently in a series of experiments on the perception of texture. The findings from these experiments are particularly important because they not only demonstrate that channels interact in the perception of tactile texture but also show that the channels that were originally discovered in the temporal domain of vibratory stimulation are operative in the spatial domain of texture perception as well. It is remarkable that knowledge of the characteristics of tactile channels first obtained through experiments in which stimuli were applied through a vibrator at a single skin site appears to be directly applicable to understanding how the roughness of a surface consisting of spatially distributed variations in texture is perceived. Indeed, it was the finding that psychophysical thresholds for detecting low- but not high-frequency vibration become elevated when the sharp spatial gradient in stimulus amplitude is eliminated by removal of the rigid surround (Gescheider et al., 1978; Verrillo, 1962, 1979b) that gave rise to the hypothesis that the receptors that detect low-frequency vibration, now known to be Merkel-neurite complexes with their SA I nerve fibers and Meissner corpuscles with their RA nerve fibers, are highly responsive to sharp gradients in stimulus intensity. Furthermore, it was concluded that these edge-detecting receptors would be ideally suited for detecting the spatially distributed variations in stimulation that characterize the textured surfaces of objects (Verrillo, 1979b).

For many years the results of psychophysical and neurophysiological experiments have strongly linked the activity of SA I nerve fibers to the perception of tactile roughness. The type of textured surface used in several of these experiments is illustrated in Figure 7.5 (Bolanowski et al., 1997; Connor et al., 1990; Gescheider, Bolanowski, Greenfield, & Brunette, 2005). The stimuli consist of raised dots arranged symmetrically on a flat surface; each dot forms a truncated cone, with sides rising at an angle of 60°. The height (h) and width (w) of each dot and the distance between dots (d) can be systematically varied. In an experiment by Connor et al. (1990), action potentials were recorded from single nerve fibers innervating the monkey's fingertip as the raised-dot pattern was presented at different positions on the fingertip. Magnitude estimates of perceived roughness were also obtained from human observers as they moved their index fingertips over the same patterns. The spatial variation in the firing rates of the nerve fiber, specified as the difference between the firing rates of the fiber as the dot pattern is moved across the skin from one position to the next, correlates highly with magnitude estimates of the roughness of raised-dot patterns (Connor et al., 1990). But the correlations between judged roughness and spatial variation in firing rate are much lower for RA and PC fibers than for SA I fibers (see Figure 7.6). However, this finding does not necessarily mean that activity in the SA I channel is the sole determinant of the percep-

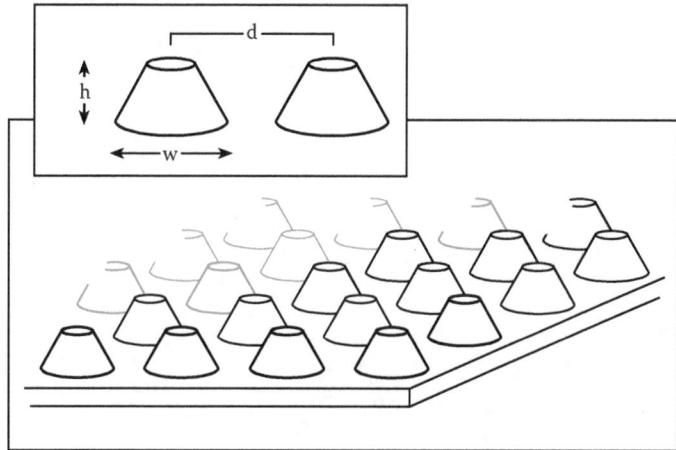

Figure 7.5 Illustration of raised-dot pattern in which *h* is dot height, *w* is dot width and *d* is interdot distance. From Gescheider, Bolanowski, Greenfield, and Brunette (2005).

tion of roughness, as some investigators have proposed (e.g., Johnson & Hsiao, 1992; Johnson, 2001).

According to the channel-interaction hypothesis, in principle it would be possible for perceived roughness to be influenced by neural activity in all four channels acting in consort. Indeed, Bolanowski and his co-workers (Bolanowski et al., 1997) reported that the perceived roughness of raised-dot patterns was significantly reduced following selective adaptation of the PC channel with a 250-Hz stimulus applied to the fingertip. This finding indicates that in the unadapted state, the perceived roughness of a dot pattern is influenced by activity in the PC channel. This conclusion was also supported by the results of a multidimensional scaling analysis in which it was found that three orthogonal perceptual dimensions could account for the judged dissimilarities of raised-dot patterns (Gescheider, Bolanowski, Greenfield, & Brunette, 2005). Selective adaptation of the PC channel changed the perceptual clarity of the pattern, which correlated highly with the judged smoothness of the individual dots. Specifically, the dots felt rougher when the fingertip was not adapted but felt smoother and clearly stood out from their background following adaptation of the PC channel. It was concluded that the PC channel processes information about the roughness of the microstructure of a textured surface, whereas the roughness of the pattern is processed by the SA I channel.

It is significant that the PC and SA I channels can combine their information about the roughness of a surface when observers attend to both the microstructure and the macrostructure of a surface. The results of an experiment in which observers were instructed to attend to the overall roughness of the

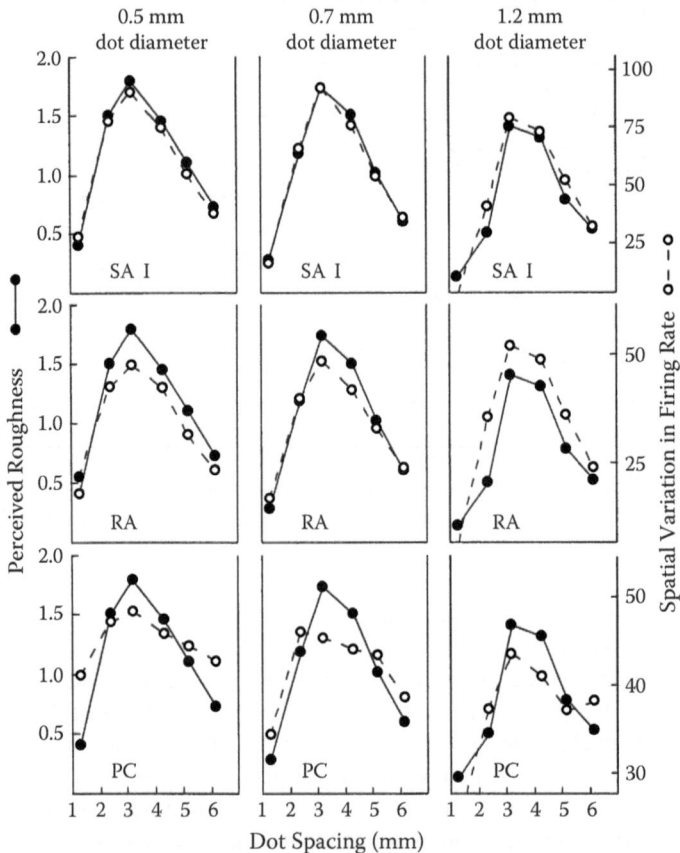

Figure 7.6   Perceived roughness judgments from human observers and spatial variation in the neural responses of the SA I, RA, and PC nerve fibers in the monkey as a
function of dot spacing and dot diameter in raised-dot patterns. From Connor et al.
(1990).

surface defined as a combination of the roughness of the pattern of dots and the
roughness of the individual dots are seen in Figure 7.7. Magnitude estimates
of overall roughness were an inverted U-shaped function of dot spacings, with
the dot pattern feeling roughest when the spacing between the raised dots was
3.5 mm. Selective adaptation of the PC channel with a 250-Hz adapting stimulus substantially reduced perceived roughness at all dot spacings except at
the lowest dot-spacing value employed in the experiment. These results are in
agreement with those reported by Bolanowski et al. (1997) and strongly suggest that the overall roughness of a textured surface is influenced by activity in
the PC channel, which appears to contribute information about the roughness
of the fine-grain elements of the surface.

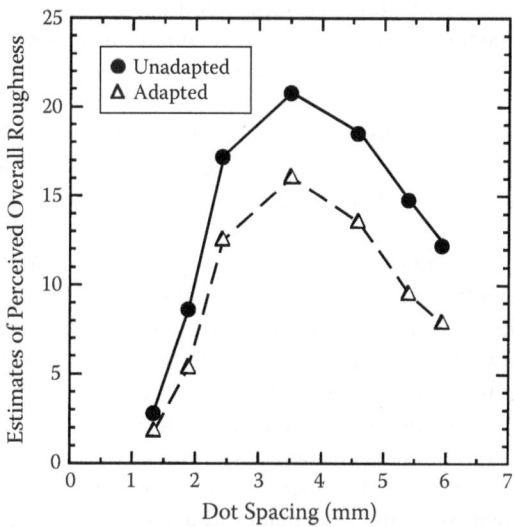

Figure 7.7 Magnitude estimates of overall roughness as a function of dot spacing before and after adaptation of the PC channel with a 250-Hz adapting stimulus. From Gescheider, Bolanowski, Greenfield, and Brunette (2005).

This conclusion is also consistent with the results of a series of experiments by Hollins and his co-workers, who have proposed that there are two separate mechanisms underlying the tactile perception of texture (Bensmaia & Hollins, 2003; Hollins et al., 1998; Hollins et al., 2001; Hollins et al., 2000; Hollins, Lorenz, & Harper, 2006; Hollins & Risner, 2000). According to their duplex theory of texture perception, coarse textures with elements larger than 0.2 mm are processed by the SA I channel, which encodes roughness through spatial variations in neural activity within a population of SA I fibers. In contrast, the perception of fine textures with elements smaller than 0.2 mm are thought to be mediated by the detection of vibratory information generated by movement of the skin over the object or movement of the object over the skin. This vibratory information is processed mainly by PC fibers. Indeed, Bensmaia and Hollins (2003) have reported that the perceived roughness of textured surfaces with elements separated by less than 0.2 mm is a function of the vibration energy, which is produced when the surface is passively moved over the finger pad, weighted by its effectiveness in stimulating the PC channel. Essentially, these investigators found that a surface felt rougher to the extent that vibrations resulting from moving a fine-textured surface over the skin were effective in stimulating the PC channel with its high-frequency tuning characteristic. Furthermore, tactile inspection with the fingertip of a textured surface can, as a result of adaptation, reduce the perceived roughness of finely but not coarsely textured surfaces (Hollins et al., 2006). This is because movement of the fingertip over the finely textured surfaces produced vibratory stimulation of sufficiently high frequency to excite the PC channel.

The results obtained by Hollins and his co-workers, Bolanowski et al. (1997), and Gescheider, Bolanowski, Greenfield, and Brunette (2005), in conjunction with those of Connor et al. (1990), clearly demonstrate that both the PC channel and the SA I channel are involved in the processing of information about the roughness of a textured surface. If the PC channel is specialized in processing fine-grain texture capable of generating vibratory cues strong enough to excite PC fibers, and the SA I channel is specialized in processing coarser textures by responding to spatial variations in the firing rates of SA I fibers, then an important question arises: do the channels interact in the perception of a textured surface, or do they make their respective contributions to the perception of roughness separately and independently?

The answer to this question appears to be that the channels interact in the perception of some textured surfaces in which both fine-grain and coarse elements are present. What has been demonstrated experimentally thus far is that both the PC channel and the SA I channel contribute to the overall perceived roughness of a raised-dot pattern in which the dots contain fine-grain elements at their tips but the pattern of the dots is coarse grain by virtue of the separation between dots being greater than 0.2 mm. This is illustrated in Figure 7.7, where it is evident that adaptation of the PC channel reduces the overall perceived roughness of a raised-dot pattern, but it must be pointed out that this effect is highly dependent upon attentional processes. Figure 7.8 shows that adaptation of the PC channel had no effect on magnitude estimates of roughness when observers were instructed to judge the roughness of the *pattern* (Gescheider, Bolanowski, Greenfield, & Brunette, 2005). In this situation it appears that the observers' perceptions of the surface are organized globally, with the elements comprising the pattern tending not to emerge as salient perceptual features. The Gestaltists argued that perceptual processing typically proceeds from a global analysis of the whole pattern to a more detailed analysis of the component parts (Kimchi, 1992; Navon, 1977; Watt, 1988), and this principle may be operating here.

Thus, the roughness of a pattern may take priority over the roughness of the individual dots making up the pattern unless the observer is specifically instructed to attend to both. Only when observers were instructed to make judgments of overall roughness defined as the roughness of the individual raised dots and the pattern of the dots did adaptation of the PC channel result in lower magnitude estimates of roughness (Figure 7.7). This observation could also explain why in some studies (e.g., Lederman, Loomis, & Williams, 1982) no effects of adaptation of the PC channel on perceived roughness are found, whereas in others (Bolanowski et al., 1997; Gescheider, Bolanowski, Greenfield, & Brunette, 2005) they are. It is our contention that the perception

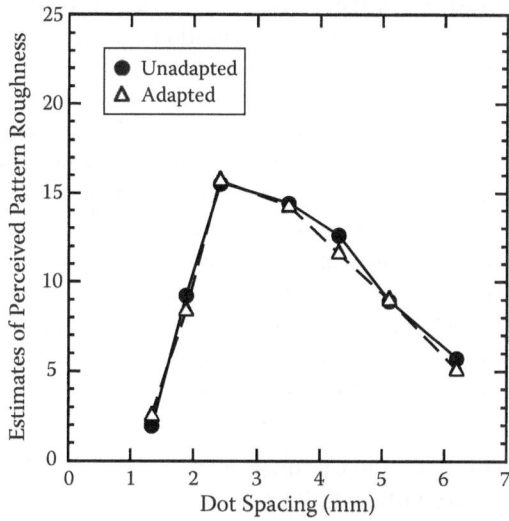

Figure 7.8 Magnitude estimates of pattern roughness as a function of dot spacing before and after adaptation of the PC channel with a 250-Hz adapting stimulus. From Gescheider, Bolanowski, Greenfield, and Brunette (2005).

of tactile stimuli results from the blended activity of multiple tactile channels. This statement must be qualified, however, by pointing out that the interaction of channels is clearly dependent upon processes of attention. Specifically, if attention is allocated entirely to the activity of one channel, then the interaction with activity in other channels will be greatly reduced.

Role of Attention in the Enhancement and Summation of Sensation Magnitude

The general principle that tactile perception results from the blending of activity in multiple channels, with the qualification that such blending often depends on whether it is possible to attend to activity in more than one channel, can also be demonstrated in changes in the perception of sensation magnitude resulting from enhancement and summation. When an observer is instructed to distribute attention over two brief stimuli presented in rapid succession, one to the PC channel and the other to the RA channel, the sensation magnitude of the pair is judged to be equal to the sum of the judged sensation magnitudes of each member of the stimulus pair (Marks, 1979; Verrillo & Gescheider, 1975). This finding clearly demonstrates that activity in different channels can be blended in the perception of the sensation magnitude of complex vibratory stimuli, provided the observer can distribute attention over the channels activated. In contrast, when the observer is instructed to attend only to the second member of the stimulus pair, its sensation magnitude is unaffected by the presentation within a different channel of the first member (Figure 4.1B). When both stimuli are presented within the same channel, the presentation of the first stimulus, whether it is attended to or not, affects the sensation magnitude of the second stimulus—an effect produced entirely by the summation of the energy in the two stimuli. Thus, interactions within channels—at least in the perception of sensation magnitude—do not appear to require the distribution of attention over the interacting stimulus components, whereas interactions between channels apparently do. These findings suggest that interactions between channels occur at a higher level in the central nervous system than do interactions within channels because the former are affected by attentional processes whereas the latter apparently are not.

Interactions Between Tactile Channels and Other Somatosensory Submodalities

Common somatosensory experiences are often the result of the combination of information conveyed by various somatosensory submodalites. For example, Weber (1846) first noted that objects of equal weight are perceived to be heavier when cold than when warm. Later, Bentley (1900), working with a wide range of cold, warm, and thermally neutral pressures, concluded that perceived wetness is the result of a combination of blends of uniform pressure and cold. The perception of oiliness can be induced by a blend of warmth and perceived pressure (Cobbey & Sullivan, 1922).

In addition to interactions between touch and temperature, interactions between touch and pain also exist. Typically these interactions are of an inhibitory nature, with touch reducing pain sensitivity (Bini, Cruccu, Hagbarth, Schady, & Torebjörk, 1984; Ekblom & Hansson, 1982; Lundberg, 1983; Melzack, Wall, & Weisz, 1963; Ottoson, Ekblom, & Hansson, 1981; Pertovaara, 1979; Sherer, Clelland, O'Sullivan, Doleys, & Canan, 1986; Sullivan, 1968; Wall & Cronly-Dillon, 1960; Zoppi, Voegelin, Signorini, & Zamponi, 1991) or pain reducing touch sensitivity (Apkarian, Stea, & Bolanowski, 1994; Bolanowski, Gescheider, Fontana, Niemiec, & Tromblay, 2001; Bolanowski, Maxfield, Gescheider, & Apkarian, 2000; Hollins, Sigurdsson, Fillingim, & Goble, 1996).

Shown in Figure 7.9 is the relationship between judgments of the sensation magnitude of 10-Hz and 100-Hz vibrotactile stimuli applied to the thenar eminence made with and without co-localized moderate heat pain induced by elevating the temperature of the skin directly under and around the 3-cm^2 area of tactile stimulation to 42–44^{0}C. In both the RA and PC channels, the perceived intensity of suprathreshold vibratory stimuli was reduced by heat-induced pain. Thus, in this case the nature of the interaction between a tactile channel and activity in another sensory modality is inhibitory. The hypothesis

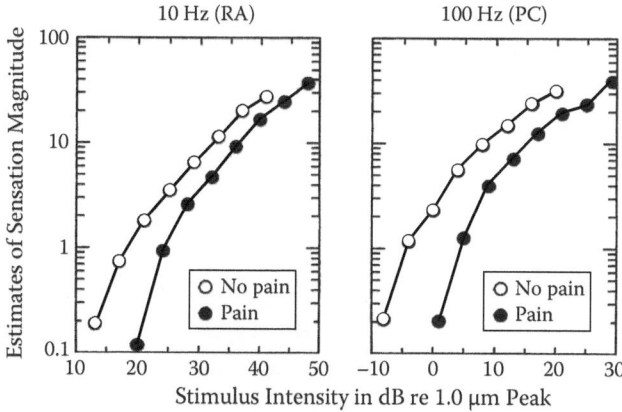

Figure 7.9 Magnitude estimates of the sensation magnitude of 10-Hz (RA channel) and 100-Hz (PC channel) stimuli as a function of stimulus intensity with and without moderate heat pain at the test site on the thenar eminence. From Bolanowski et al. (2000).

that this effect is mediated by sensory rather than attentional mechanisms is supported by the findings that auditory detection thresholds are unaffected by heat pain on the hand (Apkarian et al., 1994) and that tactile detection thresholds are unaffected by heat pain on the contralateral hand (Bolanowski et al., 2000). The principle that begins to emerge from the results of these experiments is that tactile perceptions are determined not only by the interactions of tactile channels but also by the interactions of tactile channels with activity in other somatosensory submodalities, such as temperature and pain.

8
Conclusions

The findings and analysis presented in this monograph support the conclusion that, for the glabrous skin of the hand, four separate and independent channels operate in the sense of touch. Each channel has a set of unique properties that endow it with the capacity to respond optimally, but not exclusively, to highly specific types of tactile stimulation. Each of the unique channels provides information about specific aspects of the tactile stimulus to which it is optimally responsive, and this specialization of function greatly facilitates information processing.

The PC channel, by virtue of its capacity for energy integration and its highly tuned frequency selectivity, is exceptionally sensitive to vibratory stimulation. In sharp contrast, the RA, SA I, and SA II channels are not capable of energy integration. These channels have broadly tuned frequency selectivity and consequently are much less sensitive to vibratory stimulation. Because these non-PC channels process information about the amplitude of skin displacement rather than stimulus energy, they are specialized in other important ways. The RA channel, with its relatively flat frequency-selectivity function, is able to process information about changes in the waveform of skin displacement over time. Thus, the RA channel is well suited for detecting differences in the frequency of vibratory stimulation.

In complementary fashion, the SA I channel, with its high density of receptors and associated nerve fibers with small receptive fields and its capacity for sustained activation over the entire duration of stimulation, is capable of sustained processing of information about variation in skin displacement over space. Thus, the SA I channel is well suited for discriminating differences in the spatial properties of stimuli, such as in reading Braille or in perceiving the roughness of a textured surface. The RA channel, with its high density of receptors and associated nerve fibers with small receptive fields, is also well suited for analyzing the spatial details of the stimulus, particularly if the stimulus is moving over the skin. Because of the rapidly adapting nature of the

RA channel, static tactile stimuli may be less effective in activating this channel than dynamic tactile stimuli, which move over the skin. Finally, the SA II channel, with its sparsely distributed receptors and nerve fibers that have large receptive fields, does not appear to be capable of contributing significantly to spatial processing, but instead appears to be particularly responsive to the stretching of the skin and thus may play an important role in the perception of hand conformation.

Although the tactile channels are highly specialized, it would be a serious mistake to conclude that each responds exclusively to particular dimensions of the tactile stimulus. The superior performance of a channel on a particular task does not imply non-involvement of the other channels in performing the same function in that task. For example, the finding that spatial-acuity thresholds measured either psychophysically or neurophysiologically are lowest for the SA I channel does not mean that the other three channels do not contribute to performance in spatial-acuity tasks. Instead, this finding indicates that near the limits of spatial-acuity performance where the threshold for spatial-acuity discrimination is barely exceeded, one channel—the SA I channel—exclusively mediates the discrimination of one spatially distinct stimulus from another. But when the thresholds of the other channels are exceeded, spatial-acuity performance can potentially be facilitated by activity in all of the channels. Thus, it would be a mistake, and one made by many investigators, to conclude that a neural system such as a channel is solely responsible for performance in a particular type of tactile task. A particular channel may be exclusively responsible for such performance at threshold, but once threshold has been exceeded, the other channels begin to contribute to performance in that task.

Such findings show that the world of touch, much like the world of sight and hearing, is too complex for processing by a single receptor type with simple neural connections to the brain. Instead, several independent information-processing channels are required in order for all of the potential tactile information in the external world to be sensed, processed, and blended to produce the complex tactile sensations and perceptions experienced in everyday life. The validity of this analysis of the sense of touch is demonstrated by several important findings: First, it is well established that there are several different types of tactile receptors in the skin and that each responds optimally to different types of external stimulation. Second, the specific nerve fibers associated with each of the different receptor types neurally encode the tactile stimulus exciting the receptors in very different ways to optimize the processing of the information contained in the stimulus. For example, whereas only a single neural spike in 5 to 10 RA fibers is required for stimulus detection, several spikes are required in PC fibers for stimulus detection to occur. Third, processing and encoding within a channel are not affected by activity in any of the other tactile channels, as indicated by the finding that enhancement, adaptation, masking, and sensory learning occur within but not across channels. The hypothesis that tactile channels can remain independent within the central nervous system as well as at the level of their receptors and peripheral afferent nerve fibers is supported by recordings of the neural activity in the primary somatosensory cortex of primates. Spatially distinct cortical domains are activated when stimuli are presented through the PC, RA and SA channels (Friedman, Chen, & Roe, 2004). Fourth, the channels do interact at higher levels in the central nervous system, where their individual inputs are blended to produce complex tactile perceptions. For example, roughness perception clearly involves the interaction of the PC and SA I channels, and the sensation magnitude of a stimulus is determined by the combined activity of the separate channels activated by that stimulus. Although the hypothesis that the tactile perception of suprathreshold stimuli results from the blending of activity in separate channels is supported by these findings, what remains to be discovered are the psychophysical processes and physiological mechanisms through which channel interactions occur and how these determine other forms of complex tactile perception.

Lastly, there are significant benefits of having separate channels in a sensory system such as touch. Recall that the concept of multiple channels in sensory systems evolved from the fundamental problem of determining how the nervous system is able to extract specific information about various dimensions of stimuli originating in the external world. Neural systems respond independently to different aspects of stimuli and thus serve as specific information-processing channels. Although independent in the peripheral nervous system and in the early stages of the central nervous system, channels ultimately interact by combining their information to yield unified percepts. Psychophysical and neurophysiological research has also revealed that channels facilitate the performance of a sensory system in several important ways. For example, the richness of visual, olfactory, auditory, and tactile experience is enhanced by combining information from different channels to produce a wide range of perceptual experiences. In vision, for example, hundreds of unique perceptual experiences arise from the interaction of the channels that process color, brightness, and the spatial aspects of visual stimuli. Furthermore, the sensation magnitudes of tactile and auditory stimuli are determined by the interaction of channels, and tactile texture perception involves the blending of contributions from separate channels to produce sensations of texture such as smoothness and roughness.

The operation of multiple channels in a sensory system enhances information processing by: (1) reducing the masking effects of background noise by filtering out noise not within the frequency range of a specific channel, thus making stimuli easier to detect; (2) reducing ambiguity in the interpretation of neural information resulting from sensory stimulation, thus making it easier to discriminate among stimuli; and (3) increasing the richness and variety of sensory experiences by blending the information provided by the separate channels. It is clearly evident that future advances in sensory science will be closely linked to further research on the properties of channels and their interactions in the processing of sensory information. This will include the study of how populations of receptors and neurons respond to stimuli both within and across tactile channels.

References

Apkarian, A.V., Stea, R.A., & Bolanowski, S.J. (1994). Heat-induced pain diminishes vibrotactile perception: A touch gate. *Somatosensory & Motor Research*, 11, 259–267.

Bacon, S.P., & Jesteadt, W. (1987). Effects of pure-tone forward masker duration on psychophysical measures of frequency selectivity. *Journal of the Acoustical Society of America*, 82, 1925–1932.

Barlow, H.B. (1958). Temporal and spatial summation in human vision at different background intensities. *Journal of Physiology (London)*, 141, 337–350.

Barris, M.C., Dawson, W.W., & Theiss, C.L. (1980). The visual sensitivity of women during the menstrual cycle. *Documenta Opthalmologica*, 49, 293–301.

Bensmaia, S.J., & Hollins, M. (2003). The vibrations of texture. *Somatosensory and Motor Research*, 20, 33–43.

Bentley, I.M. (1900). The synthetic experiment. *American Journal of Psychology*, 11, 405–425.

Berglund, U., & Berglund, B. (1970). Adaption and recovery in vibrotactile perception. *Perceptual and Motor Skills*, 30, 843–853.

Békésy, G. von (1939). Über die Vibrationsempfindung. *Akustische Zeitschrift*, 4, 316–334.

Bini, G., Cruccu, G., Hagbarth, K.-E., Schady, W., & Torebjörk, E. (1984). Analgesic effect of vibration and cooling on pain induced by intraneural electrical stimulation. *Pain*, 18, 239–248.

Blakemore, C., & Campbell, F.W. (1969). On the existence of neurones in the human visual system selectively sensitive to the orientation and size of retinal images. *Journal of Physiology (London)*, 203, 237–260.

Bolanowski, S.J., Jr., (1981). Intensity and frequency characteristics of Pacinian corpuscles. Ph.D. Dissertation and Special Report, ISR-S-20, Institute for Sensory Research, Syracuse University, Syracuse, NY.

Bolanowski, S.J., Cohen, J.C., & Pawson, L. (1997). The Pacinian (P), as well as the non-Pacinian (NP) channels, contributes to the perception of roughness. *Society for Neuroscience Abstracts*, 23, 1003.

Bolanowski, S.J., Gescheider, G.A., Fontana, A.M., Niemiec, J.L., & Tromblay, J.L. (2001). The effects of heat-induced pain on the detectability, discriminability, and sensation magnitude of vibrotactile stimuli. *Somatosensory & Motor Research*, 18, 5–9.

Bolanowski, S.J., Gescheider, G.A., & Verrillo, R.T. (1994). Expansion of the four-channel model for touch: Hairy skin. *Somatosensory & Motor Research*, 11, 279–290.

Bolanowski, S.J., Jr., Gescheider, G.A., Verrillo, R.T., & Checkosky, C.M. (1988). Four channels mediate the mechanical aspects of touch. *Journal of the Acoustical Society of America*, 84, 1680–1694.

Bolanowski, S.J., Hall, K.L., Makous, J.C., & Merzenich, M.M. (1995). Intensity discrimination of vibratory stimuli can be improved by training. *Society for Neuroscience Abstracts*, 21, 1443.

Bolanowski, S.J., Maxfield, L.M., Gescheider, G.A., & Apkarian, A.V. (2000). The effects of stimulus location on the gating of touch by heat- and cold-induced pain. *Somatosensory & Motor Research*, 17, 195–204.

Bolanowski, S.J., Jr., & Verrillo, R.T. (1982). Temperature and criterion effects in a somatosensory subsystem: A neurophysiological and psychophysical study. *Journal of Neurophysiology*, 48, 836–855.

Buck, L.B. (1996). Information coding in the vertebrate olfactory system. *Annual Review of Neuroscience,* 19, 517–554.

Buck, L.B., & Axel, R. (1991). A novel multigene family may encode odorant receptors: A molecular basis for odor recognition. *Cell*, 65, 175–187.

Campbell, F.W., & Robson, J.G. (1968). Application of Fourier analysis to the visibility of gratings. *Journal of Physiology (London)*, 197, 551–566.

Capraro, A.J., Verrillo, R.T., & Zwislocki, J.J. (1979). Psychophysical evidence for a triplex system of cutaneous mechanoreception. *Sensory Processes,* 3, 334–352.

Cauna, N. (1965). The effects of aging on the receptor organs of the human dermis. In W. Montagna (Ed.), *Advances in biology of the skin: Aging* (Vol. VI, pp. 63–96). New York: Pergamon Press.

Chambers, M.R., Andres, K.H., von Düering, M., & Iggo, A. (1972). The structure and function of the slowly adapting type II mechanoreceptor in hairy skin. *Quarterly Journal of Experimental Physiology and Cognate Medical Sciences,* 57, 417–445.

Checkosky, C.M., & Bolanowski, S.J. (1992). The effects of stimulus duration on the response properties of Pacinian corpuscles: Implications for the neural code. *Journal of the Acoustical Society of America*, 91, 3372–3380.

Cobbey, L.W., & Sullivan, A.H. (1922). An experimental study of the perception of oiliness. *American Journal of Psychology*, 33, 121–127.

Cohen, L.H., & Lindley, S.B. (1938). Studies in vibratory sensibility. *American Journal of Psychology,* 51, 44–63.

Connor, C.E., Hsiao, S.S., Phillips, J.R., & Johnson, K.O. (1990). Tactile roughness: Neural codes that account for psychophysical magnitude estimates. *Journal of Neurophysiology*, 10, 3823–3836.

Connor, C.E., & Johnson, K.O. (1992). Neural coding of tactile texture: Comparison of spatial and temporal mechanisms for roughness perception. *Journal of Neurophysiology*, 12, 3414–3426.

Cornsweet, T.N. (1970). *Visual perception* (pp. 96–105). New York: Academic Press.

Craig, J.C., & Lyle, K.B. (2001). A comparison of tactile spatial sensitivity on the palm and fingerpad. *Perception & Psychophysics*, 63, 337–347.

Craig, J.C., & Lyle, K.B. (2002). A correction and a comment on Craig and Lyle (2001). *Perception & Psychophysics*, 64, 504–506.

Darian-Smith, I., & Kenins, P. (1980). Innervation density of mechanoreceptive fibres supplying glabrous skin of the monkey's index finger. *Journal of Physiology (London)*, 309, 147–155.

Darian-Smith, I., & Oke, L.E. (1980). Peripheral neural representation of the spatial frequency of a grating moving across the monkey's fingerpad. *Journal of Physiology (London)*, 309, 117–133.

Dellon, A.S. (1981). *Evaluation of sensibility and re-education of sensation in the hand*. Baltimore: Williams and Wilkins.

DeValois, R.L., & DeValois, K.K. (1975). Neural coding of color. In E.C. Carterette & M.P. Friedman (Eds.), *Handbook of perception* (Vol. 5, pp. 117–166). New York: Academic Press.

Ekblom, A., & Hansson, P. (1982). Effects of conditioning vibratory stimulation on pain threshold of the human tooth. *Acta Physiologica Scandinavia*, 114, 601–604.

Erickson, R.P. (1968). Stimulus coding in topographic and nontopographic afferent modalities: On the significance of the activity of individual sensory neurons. *Psychological Review*, 75, 447–465.

Erickson, R.P. (1978). Common properties of sensory systems. In R.B. Masterton (Ed.), *Handbook of behavioral neurobiology* (Vol. 1, pp. 73–90). New York: Plenum Press.

Erickson, R.P. (1982). The across-fiber pattern theory: An organizing principle for molar neural function. In W.D. Neff (Ed.), *Contributions to sensory physiology* (Vol. 6, pp. 79–110). New York: Academic Press.

Erickson, R.P., & Schiffman, S.S. (1975). The chemical senses: A systematic approach. In M.S. Gazzaniga & C. Blakemore (Eds.), *Handbook of psychobiology* (pp. 393–426). New York: Academic Press.

Essick, G.K. (1998). Factors affecting direction discrimination of moving tactile stimuli. In J.W. Morley (Ed.), *Neural aspects in tactile sensation* (pp. 1–54). Amsterdam: Elsevier Press.

Feldtkeller, R., & Zwicker, E. (1956). *Das Ohr als Nachrichtenempfänger*. Stuttgart: S. Hirzel Verlag.

Fletcher, H. (1940). Auditory patterns. *Review of Modern Physics*, 12, 47–65.

Florentine, M., Buus, S., Scharf, B., & Zwicker, E. (1980). Frequency selectivity in normally-hearing and hearing-impaired observers. *Journal of Speech and Hearing Research*, 23, 646–669.

Friedman, R.M., Chen, L.M., & Roe, A.W. (2004). Modality maps within primate somatosensory cortex. *Proceedings of the National Academy of Science*, 101, 12724–12729.

Frisina, R.D., & Gescheider, G.A. (1977). Comparison of child and adult vibrotactile thresholds as a function of frequency and duration. *Perception & Psychophysics,* 22, 100–103.

Garner, W.R., & Miller, G.A. (1947). The masked threshold of pure tones as a function of duration. *Journal of Experimental Psychology,* 37, 293–303.

Gässler, G. (1954). Über die Hörschwelle für Schallereignisse mit Verschieden breitem Frequenz-spektrum. *Acustica,* 4, 408–414.

Gescheider, G.A. (1976). Evidence in support of the duplex theory of mechanoreception. *Sensory Processes,* 1, 68–76.

Gescheider, G.A. (1997). *Psychophysics: The fundamentals* (3rd ed., pp. 31–44). Mahwah, New Jersey: Lawrence Erlbaum Associates.

Gescheider, G.A., Beiles, E.J., Bolanowski, S.J., Checkosky, C.M., & Verrillo, R.T. (1994). The effects of aging on information-processing channels in the sense of touch: Temporal summation in the P channel. *Somatosensory & Motor Research,* 11, 359–369.

Gescheider, G.A., Berryhill, M.E., Verrillo, R.T., & Bolanowski, S.J. (1999). Vibrotactile temporal summation: Probability summation or neural integration? *Somatosensory & Motor Research,* 16, 229–242.

Gescheider, G.A., Bolanowski, S.J., & Chatterton, S.K. (2003). Temporal gap detection in tactile channels. *Somatosensory & Motor Research,* 20, 239–247.

Gescheider, G.A., Bolanowski, S.J., Greenfield, T.C., & Brunette, K. (2005). Perception of the tactile texture of raised-dot patterns: A multidimensional analysis. *Somatosensory & Motor Research,* 22, 127–140.

Gescheider, G.A., Bolanowski, S.J., Hall, K.L., Hoffman, K.E., & Verrillo, R.T. (1994). The effects of aging on information-processing channels in the sense of touch: I. Absolute sensitivity. *Somatosensory & Motor Research,* 11, 345–357.

Gescheider, G.A., Bolanowski, S.J., & Hardick, K.R. (2001). The frequency selectivity of information-processing channels in the tactile sensory system. *Somatosensory & Motor Research,* 18, 191–201.

Gescheider, G.A., Bolanowski, S.J., Pope, J.V., & Verrillo, R.T. (2002). A four-channel analysis of the tactile sensitivity of the fingertip: Frequency selectivity, spatial summation, and temporal summation. *Somatosensory & Motor Research,* 19, 114–124.

Gescheider, G.A., Bolanowski, S.J., & Verrillo, R.T. (1989). Vibrotactile masking: Effects of stimulus-onset asynchrony and stimulus frequency. *Journal of the Acoustical Society of America,* 85, 2059–2064.

Gescheider, G.A., Bolanowski, S.J., & Verrillo, R.T. (2004). Some characteristics of tactile channels. *Behavioural Brain Research,* 148, 35–40.

Gescheider, G.A., Capraro, A.J., Frisina, R.D., Hamer, R.D., & Verrillo, R.T. (1978). The effects of a rigid surround on vibrotactile thresholds. *Sensory Processes,* 2, 99–115.

Gescheider, G.A., Frisina, R.D., & Verrillo, R.T. (1979). Selective adaptation of vibrotactile thresholds. *Sensory Processes,* 3, 37–48.

Gescheider, G.A., Güçlü, B., Sexton, J.L., Karalunas, S., & Fontana, A. (2005). Spatial summation in the tactile sensory system: Probability summation and neural integration. *Somatosensory & Motor Research,* 22, 255–268.

Gescheider, G.A., Hoffman, K.E., Harrison, M.A., Travis, M.L., & Bolanowski, S.J. (1994). The effects of masking on vibrotactile temporal summation in the detection of sinusoidal and noise stimuli. *Journal of the Acoustical Society of America*, 95, 1006–1016.

Gescheider, G.A., & Joelson, J.M. (1983). Vibrotactile temporal summation for threshold and suprathreshold levels of stimulation. *Perception & Psychophysics*, 33, 156–162.

Gescheider, G.A., & Migel, N. (1995). Some temporal parameters in vibrotactile forward masking. *Journal of the Acoustical Society of America*, 98, 3195–3199.

Gescheider, G.A., O'Malley, M.J., & Verrillo, R.T. (1983). Vibrotactile forward masking: Evidence for channel independence. *Journal of the Acoustical Society of America*, 74, 474–485.

Gescheider, G.A., Santoro, K.E., Makous, J.C., & Bolanowski, S.J. (1995). Vibrotactile forward masking: Effects of the amplitude level and duration of the masking stimulus. *Journal of the Acoustical Society of America*, 98, 3188–3194.

Gescheider, G.A., Sklar, B.F., Van Doren, C.L., & Verrillo, R.T. (1985). Vibrotactile forward masking: Psychophysical evidence for a triplex theory of cutaneous mechanoreception. *Journal of the Acoustical Society of America*, 78, 534–543.

Gescheider, G.A., Valetutti, A.A., Jr., Padula, M.C., & Verrillo, R.T. (1992). Vibrotactile forward masking as a function of age. *Journal of the Acoustical Society of America*, 91, 1690–1696.

Gescheider, G.A., & Verrillo, R.T. (1979). Vibrotactile frequency characteristics determined by adaptation and masking procedures. In D.R. Kenshalo (Ed.), *Sensory functions of the skin of humans* (pp. 183–205). New York: Plenum Press.

Gescheider, G.A., & Verrillo, R.T. (1982). Contralateral enhancement and suppression of vibrotactile sensation. *Perception & Psychophysics*, 32, 69–74.

Gescheider, G.A., Verrillo, R.T., Capraro, A.J., & Hamer, R.D. (1977). Enhancement of vibrotactile sensation magnitude and predictions from the duplex model of mechanoreception. *Sensory Processes*, 1, 187–203.

Gescheider, G.A., Verrillo, R.T., McCann, J.T., & Aldrich, E.M. (1984). Effects of the menstrual cycle on vibrotactile sensitivity. *Perception & Psychophysics*, 36, 586–592.

Gescheider, G.A., Verrillo, R.T., & Van Doren, C.L. (1982). Prediction of vibrotactile masking functions. *Journal of the Acoustical Society of America*, 72, 1421–1426.

Gescheider, G.A., & Wright, J.H. (1968). Effects of sensory adaptation on the form of the psychophysical magnitude function for cutaneous vibration. *Journal of Experimental Psychology*, 77, 308–313.

Gescheider, G.A., & Wright, J.H. (1969). Effects of vibrotactile adaptation on the perception of stimuli of varied intensity. *Journal of Experimental Psychology*, 81, 449–453.

Gescheider, G.A., Zwislocki, J.J., & Rasmussen A. (1996). Effects of stimulus duration on the amplitude DL for vibrotaction. *Journal of the Acoustical Society of America*, 100, 2312–2319.

Goble, A.K., & Hollins, M. (1993). Vibrotactile adaptation enhances amplitude discrimination. *Journal of the Acoustical Society of America*, 93, 418–424.

Goodwin, A.W., Browning, A.S., & Wheat, H.E. (1995). Representation of curved surfaces in responses of mechanoreceptive afferent fibers innervating the monkey's fingerpad. *Journal of Neuroscience,* 15, 798–810.

Goodwin, A.W., & Wheat, H.E. (1992). Magnitude estimation of contact force when objects with different shapes are applied passively to the fingerpad. *Somatosensory & Motor Research,* 9, 339–344.

Goodwin, A.W., & Wheat, H.E. (2004). Sensory signals in neural populations underlying tactile perception and manipulation. *Annual Review of Neuroscience,* 27, 53–77.

Goolkasian, P. (1980). Cyclic changes in pain perception: An ROC analysis. *Perception & Psychophysics,* 27, 499–504.

Graham, C.H., Brown, R.H., & Mote, F.A., Jr. (1939). The relation of size of stimulus and intensity in the human eye: I. Intensity thresholds for white light. *Journal of Experimental Psychology,* 24, 555–573.

Graham, C.H., & Margaria, R. (1935). Area and the intensity-time relation in the peripheral retina. *American Journal of Physiology,* 113, 299–305.

Green, B.G. (1976). Vibrotactile temporal summation: Effect of frequency. *Sensory Processes,* 1, 138–149.

Greenwood, D.D. (1961). Auditory masking and the critical band. *Journal of the Acoustical Society of America,* 33, 484–502.

Güçlü, B., & Bolanowski, S.J. (2002). Modeling population responses of rapidly-adapting mechanoreceptive fibers. *Journal of Computational Neuroscience,* 12, 201–218.

Güçlü, B., & Bolanowski, S.J. (2004). Probability of stimulus detection in a model population of rapidly adapting fibers. *Neural Computation,* 16, 39–58.

Güçlü, B., Gescheider, G.A., Bolanowski, S.J., & Istefanopulos, Y. (2005). Population-response model for vibrotactile spatial summation. *Somatosensory & Motor Research,* 22, 239–253.

Hahn, J.F. (1966). Vibrotactile adaptation and recovery measured by two methods. *Journal of Experimental Psychology,* 71, 655–658.

Hahn, J.F. (1968). Low-frequency vibrotactile adaptation. *Journal of Experimental Psychology,* 78, 655–659.

Hamer, D.R., Verrillo, R.T., & Zwislocki, J.J. (1983). Vibrotactile masking of Pacinian and non-Pacinian channels. *Journal of the Acoustical Society of America,* 73, 1293–1303.

Hecht, S., Shlaer, S., & Pirenne, M.H. (1942). Energy, quanta, and vision. *Journal of General Physiology,* 25, 819–840.

Henkin, R.I. (1974). Sensory changes during the menstrual cycle. In M. Ferin, F. Halberg, R.M. Richart, & R.L. van de Wiele (Eds.), *Biorhythms and human reproduction.* New York: Wiley.

Henning, H. (1916a). *Der Geruch.* Leipzig: Barth.

Henning, H. (1916b). Die Qualitätenreihe des Geschmacks. *Zeitschrift für Psychologie,* 74, 203–219.

Hollins, M., Bensmaia, S., & Risner, R. (1998). The duplex theory of tactile texture perception. In S. Grondin & Y. Lacouture (Eds.), *Fechner Day 98: Proceedings of the fourteenth annual meeting of the international society for psychophysics* (pp. 115–120). Quebec, Canada.

Hollins, M., Bensmaia, S.J., & Washburn, S. (2001). Vibrotactile adaptation impairs discrimination of fine, but not coarse, textures. *Somatosensory & Motor Research,* 18, 253–262.

Hollins, M., Delemos, K.A., & Goble, A.K. (1991). Vibrotactile adaptation on the face. *Perception & Psychophysics*, 49, 21–30.

Hollins, M., Delemos, K.A., & Goble, A.K. (1996). Vibrotactile adaptation of the RA system: A psychophysical analysis. In O. Franzén, R. Johansson, & L. Terenius (Eds.), *Somesthesis and the neurobiology of the somatosensory cortex* (pp. 101–111). Basel: Birkhäuser.

Hollins, M., Fox, A., & Bishop, C. (2000). Imposed vibration influences perceived tactile smoothness. *Perception*, 29, 1455–1465.

Hollins, M., Goble, A.K., Whitsel, B.L., & Tommerdahl, M. (1990). Time course and actions spectrum of vibrotactile adaptation. *Somatosensory & Motor Research*, 7, 205–221.

Hollins, M., Lorenz, F., & Harper, D. (2006). Somatosensory coding of roughness: The effect of texture adaptation in direct and indirect touch. *Journal of Neuroscience,* 26, 5582–5588.

Hollins, M., & Risner, S.R. (2000). Evidence for the duplex theory of tactile texture perception. *Perception & Psychophysics*, 62, 695–705.

Hollins, M., & Roy, E.A. (1996). Perceived intensity of vibrotactile stimuli: The role of mechanoreceptive channels. *Somatosensory & Motor Research*, 13, 273–286.

Hollins, M., Sigurdsson, L., Fillingim, L., & Goble, A.K. (1996). Vibrotactile threshold is elevated in temporomandibular disorders. *Pain*, 67, 89–96.

Hurvich, L.M. (1981). *Color vision.* Sunderland, MA: Sinauer Associates.

Iggo, A. (1963). An electrophysiological analysis of afferent fibers in primate skin. *Acta Neurovegetativa,* 24, 225–240.

Iggo, A., & Muir, A.R. (1969). The structure and function of a slowly adapting touch corpuscle in hairy skin. *Journal of Physiology (London),* 200, 763–796.

Iggo, A., & Ogawa, H. (1977). Correlative physiological and morphological studies of rapidly adapting mechanoreceptors in the cat's glabrous skin. *Journal of Physiology (London),* 266, 275–296.

Jänig, W., Schmidt, R.F., & Zimmermann, M. (1968). Single unit responses and the total afferent outflow from the cat's foot pad upon mechanical stimulation. *Experimental Brain Research,* 6, 100–115.

Jeffress, L.A. (1975). Masking of tone by tone as a function of duration. *Journal of the Acoustical Society of America,* 58, 399–403.

Johansson, R.S. (1976). Receptive field sensitivity profile of mechanosensitive units innervating the glabrous skin of the human hand. *Brain Research,* 104, 330–334.

Johansson, R.S. (1978). Tactile sensibility in the human hand: Receptive field characteristics of mechanoreceptive units in the glabrous skin area. *Journal of Physiology (London),* 281, 101–123.

Johansson, R.S., Landström, U., & Lundström, R. (1982a). Responses of mechanoreceptive afferent units in the glabrous skin of the human hand to sinusoidal skin displacements. *Brain Research,* 244, 17–25.

Johansson, R.S., Landström, U., & Lundström, R. (1982b). Sensitivity to edges of mechanoreceptive afferent units innervating the glabrous skin of the human hand. *Brain Research,* 244, 27–35.

Johansson, R.S., & Vallbo, A.B. (1979). Tactile sensitivity in the human hand: Relative and absolute densities of four types of mechanoreceptive units in glabrous skin. *Journal of Physiology (London),* 286, 283–300.

Johnson, K.O. (2001). The roles and functions of cutaneous mechanoreceptors. *Current Opinion in Neurobiology,* 11, 445–461.

Johnson, K.O., & Hsiao, S.S. (1992). Neural mechanisms of tactual form and texture perception. *Annual Review of Neuroscience,* 15, 227–250.

Johnson, K.O., & Phillips, J.R. (1981). Tactile spatial resolution: I. Two-point discrimination, gap detection, grating resolution, and letter recognition. *Journal of Neurophysiology,* 46, 1177–1191.

Johnson, K.O., Yoshioka, T., & Vega-Bermudez, F. (2000). Tactile functions of mechanoreceptive afferents innervating the hand. *Journal of Clinical Neurophysiology,* 17, 539–558.

Karn, H.W. (1936). Area and the intensity-time relation in the fovea. *Journal of General Psychology,* 14, 360–369.

Kelly, D.H. (1961). Visual response to time-dependent stimuli: I. Amplitude sensitivity measurements. *Journal of the Optical Society of America,* 51, 422–429.

Kenshalo, D.R. (1966). The cool threshold associated with phases of the menstrual cycle. *Journal of Applied Physiology,* 21, 1031–1039.

Kenshalo, D.R. (1979). Aging effects on cutaneous and kinesthetic sensibilities. In S.S. Han & D.H. Coon (Eds.), *Special senses in aging.* Ann Arbor: Academic Press.

Kimchi, R. (1992). Primacy of wholistic processing and global/local paradigm: A critical review. *Psychological Bulletin,* 112, 24–38.

Labs, S.M., Gescheider, G.A., Fay, R.R., & Lyons, C.H. (1978). Psychophysical tuning curves in vibrotaction. *Sensory Processes,* 2, 231–247.

Lamb, G.D. (1983). Tactile discrimination of textured surfaces: Peripheral neural coding in the monkey. *Journal of Physiology (London),* 338, 567–587.

LaMotte, R.H., & Mountcastle, V.B. (1975). Capacities of humans and monkeys to discriminate between vibratory stimuli of different frequency and amplitude: A correlation between neural events and psychophysical measurements. *Journal of Neurophysiology,* 38, 539–559.

LaMotte, R.H., & Srinivasan, M.A. (1987). Tactile discrimination of shape: Responses of rapidly adapting mechanoreceptive afferents to a step stroked across the monkey fingerpad. *Journal of Neuroscience,* 7, 1672–1681.

Lederman, S.J., Loomis, J.M., & Williams, D.A. (1982). The role of vibration in the tactile perception of roughness. *Perception & Psychophysics,* 32, 109–116.

Lennie, P., & D'Zmura, M. (1988). Mechanisms of color vision. *CRC Critical Review of Neurobiology,* 3, 333–400.

Leung, Y.Y., Hsiao, S.S., & Johnson, K.O. (1994). Adaptation of mechanoreceptive afferents to continuous sinusoidal vibration. *Society for Neuroscience Abstracts,* 20, 1381.

Lindblom, U. (1965). Properties of touch receptors in distal glabrous skin of the monkey. *Journal of Neurophysiology, 28,* 966–985.

Lindblom, U., & Lund, L. (1966). The discharge from vibration-sensitive receptors in the monkey foot. *Experimental Neurology,* 15, 401–417.

Loomis, J.M. (1979). An investigation of tactile hyperacuity. *Sensory Processes,* 3, 289–302.

Loomis, J.M., & Collins, C.C. (1978). Sensitivity to shifts of a point stimulus: An instance of tactile hyperactivity. *Perception & Psychophysics,* 24, 487–492.

Loewenstein, W.R., & Mendelson, M. (1965). Components of receptor adaptation in a Pacinian corpuscle. *Journal of Physiology (London),* 177, 377–397.

Loewenstein, W.R., & Skalak, R. (1966). Mechanical transmission in a Pacinian corpuscle: An analysis and a theory. *Journal of Physiology (London),* 182, 346–378.

Lundberg, T. (1983). Vibratory stimulation for the alleviation of chronic pain. *Acta Physiologica Scandinavia,* 523, 1–51.

Lundström, R., & Johansson, R.S. (1986). Acute impairment of the sensitivity of skin mechanoreceptive units caused by vibration exposure of the hand. *Ergonomics,* 29, 687–698.

Mair, R.G., Bouffard, J.A., Engen, T., & Morton, T.H. (1978). Olfactory sensitivity during the menstrual cycle. *Sensory Processes,* 2, 90–98.

Makous, J.C., Friedman, R.M., & Vierck, C.J., Jr. (1995). A critical band filter in touch. *Journal of Neuroscience,* 15, 2808–2818.

Makous, J.C., Gescheider, G.A., & Bolanowski, S.J. (1996). Decay in the effect of vibrotactile masking. *Journal of the Acoustical Society of America,* 99, 1124–1129.

Marks, L.E. (1978). *The unity of the senses: Interrelations among the modalities.* New York: Academic Press.

Marks, L.E. (1979). Summation of vibrotactile intensity: An analog to auditory critical bands. *Sensory Processes,* 3, 188–203.

Melzack, R., Wall, P.D., & Weisz, A.Z. (1963). Masking and metacontrast phenomenon in skin sensory systems. *Experimental Neurology,* 8, 35–46.

Mountcastle, V.B., LaMotte, R.H., & Carli, G. (1972). Detection thresholds for stimuli in humans and monkeys: Comparison with threshold events in mechanoreceptive afferent nerve fibers innervating the monkey hand. *Journal of Neurophysiology,* 35, 122–136.

Mountcastle, V.B., Talbot, W.H., & Kornhuber, H.H. (1966). The neural transformation of mechanical stimuli delivered to the monkey's hand. In A.V.S. de Reuck & J. Knight (Eds.), *Touch, heat, pain and itch* (pp. 325–351). London: Churchill Livingstone.

Müller J. (1826). *Zur vergleichenden Physiologie des Gesichtssinnes des Menschen und der Thiere.* Leipzig: C. Cnobloch.

Navon, D. (1977). Forest before trees: The precedence of global features in visual perception. *Cognitive Psychology,* 9, 353–383.

Ochoa, J.L., & Torebjörk, H.E. (1983). Sensations evoked by intraneural microstimulation of single mechanoreceptor units innervating the human hand. *Journal of Physiology (London)*, 342, 633–653.

O'Mara, S., Rowe, M.J., & Tarvin, R.P. (1988). Neural mechanisms in vibrotactile adaptation. *Journal of Neurophysiology*, 59, 607–622.

Ottoson, D., Ekblom, A., & Hansson, P. (1981). Vibratory stimulation for the relief of pain of dental origin. *Pain*, 10, 37–45.

Paré, M., Behets, C., & Cornu, O. (2003). Paucity of presumptive Ruffini corpuscles in the index finger pad of humans. *Journal of Comparative Neurology*, 456, 260–266.

Paré, M., Smith, A.M., & Rice, F.L. (2002). Distribution and terminal arborizations of cutaneous mechanoreceptors in the glabrous finger pads of the monkey. *Journal of Comparative Neurology*, 445, 347–359.

Parlee, M.B. (1983). Menstrual rhythms in sensory processes: A review of fluctuations in vision, olfaction, audition, taste, and touch. *Psychological Bulletin*, 93, 539–548.

Pertovaara, A. (1979). Modification of human pain threshold by specific tactile receptors. *Acta Physiologica Scandinavia*, 107, 339–341.

Phillips, J.R., Johansson, R.S., & Johnson, K.O. (1992). Responses of human mechanoreceptive afferents to embossed dot arrays scanned across fingerpad skin. *Journal of Neuroscience*, 12, 827–839.

Phillips, J.R., & Johnson, K.O. (1981). Tactile spatial resolution. III. A continuum mechanics model of skin predicting mechanoreceptor responses to bars, edges, and gratings. *Journal of Neurophysiology*, 46, 1204–1225.

Raab, D.H., Osman, E., & Rich, E. (1963). Effects of waveform correlation and signal duration on detection of noise bursts in continuous noise. *Journal of the Acoustical Society of America*, 35, 1942–1946.

Robinson, J.E., & Short, R.V. (1977). Changes in breast sensitivity at puberty, during the menstrual cycle, and at parturition. *British Medical Journal*, 1, 1188–1191.

Rogers, C.H. (1970). Choice of stimulator frequency for tactile arrays. *IEEE transactions on man-machine systems*, MMS-11, 5–11.

Rowe, M.J. (2002). Synaptic transmission between single tactile and kinaesthetic sensory nerve fibers and their central target neurones. *Behavioural Brain Research*, 135, 197–212.

Rushton, A.H. (1972). Pigments and signals in vision. *Journal of Physiology (London)*, 220, 1–31.

Sato, M. (1961). Response of Pacinian corpuscles to sinusoidal vibration. *Journal of Physiology (London)*, 159, 391–409.

Scharf, B. (1961). Complex sounds and critical bands. *Psychological Bulletin*, 58, 205–217.

Scharf, B. (1970). Loudness and frequency selectivity at short durations. In R. Plomp & G.F. Smoorenburg (Eds.), *Frequency analysis and periodicity detection in hearing*. Leiden: Sijthoff.

Sherer, C.L., Clelland, J.A., O'Sullivan, P., Doleys, D.M., & Canan, B. (1986). The effect of two sites of high frequency vibration on cutaneous pain threshold. *Pain*, 25, 133–138.

Sherrick, C.E., Cholewiak, R.W., & Collins, A.A. (1990). The localization of low- and high-frequency vibrotactile stimuli. *Journal of the Acoustical Society of America,* 88, 169–179.

Stevens, J.C., & Choo, K.K. (1996). Spatial acuity of the body surface over the life span. *Somatosensory & Motor Research,* 13, 153–166.

Sullivan, R. (1968). Effect of different frequencies of vibration on pain-threshold detection. *Experimental Neurology,* 20, 135–142.

Talbot, W.H., Darien-Smith, I., Kornhuber, H.H., & Mountcastle, V.B. (1968). The sense of flutter vibration: Comparison of the human capacity with response patterns of mechanoreceptive afferents from the monkey hand. *Journal of Neurophysiology,* 31, 301–334.

Torebjörk, H.E., & Ochoa, J.L. (1980). Specific sensations evoked by activity in single identified sensory units in man. *Acta Physiologica Scandanavia,* 110, 445–447.

Vallbo, A.B. (1981). Sensations evoked from the glabrous skin of the human hand by electrical stimulation of unitary mechanosensitive afferents. *Brain Research,* 215, 359–363.

Vallbo, A.B., & Johansson, R.S. (1976). Skin mechanoreceptors in the human hand: Neural and psychophysical thresholds. In Y. Zotterman (Ed.), *Sensory functions of the skin of primates, with special reference to man* (pp. 185–198). New York: Pergamon Press.

Vallbo, A.B., & Johansson, R.S. (1984). Properties of cutaneous mechanoreceptors in the human hand related to touch sensation. *Human Neurobiology,* 3, 3–14.

Van Doren, C.L. (1985). Temporal summation by Pacinian corpuscles precludes entrainment at the detection threshold. *Journal of the Acoustical Society of America,* 77, 2188–2189.

Verrillo, R.T. (1962). Investigation of some parameters of the cutaneous threshold for vibration. *Journal of the Acoustical Society of America,* 34, 1768–1773.

Verrillo, R.T. (1963). Effect of contactor area on the vibrotactile threshold. *Journal of the Acoustical Society of America,* 35, 1962–1966.

Verrillo, R.T. (1965). Temporal summation in vibrotactile sensitivity. *Journal of the Acoustical Society of America,* 37, 843–846.

Verrillo, R.T. (1966). Vibrotactile sensitivity and the frequency response of the Pacinian corpuscle. *Journal of the Psychonomic Society,* 4, 135–136.

Verrillo, R.T. (1968). A duplex mechanism of mechanoreception. In D. R. Kensalo (Ed.), *The skin senses* (pp. 135–159). Springfield: Thomas.

Verrillo, R.T. (1979a). Changes in vibrotactile thresholds as a function of age. *Sensory Processes,* 3, 49–59.

Verrillo, R.T. (1979b). The effect of surface gradients on vibrotactile thresholds. *Sensory Processes,* 3, 27–36.

Verrillo, R.T. (1982). Effects of aging on the suprathreshold responses to vibration. *Perception & Psychophysics,* 32, 61–68.

Verrillo, R.T., Fraioli, A.J., & Smith, R.L. (1969). Sensory magnitude of vibrotactile stimuli. *Perception & Psychophysics,* 6, 366–372.

Verrillo, R.T., & Gescheider, G.A. (1975). Enhancement and summation in the perception of two successive vibrotactile stimuli. *Perception & Psychophysics,* 18, 128–136.

Verrillo, R.T., & Gescheider, G.A. (1977). Effect of prior stimulation on vibrotactile thresholds. *Sensory Processes*, 1, 292–300.

Viemeister, N.F., & Wakefield, G.H. (1991). Temporal integration and multiple looks. *Journal of the Acoustical Society of America*, 90, 858–865.

Vierck, C.J. (1979). Comparison of punctate, edge and surface stimulation of peripheral, slowly-adapting, cutaneous afferent units of cats. *Brain Research*, 175, 155–159.

Vogten, L.L.M. (1974). Pure-tone masking: A new result from a new method. In E. Zwicker & E. Terhardt (Eds.), *Facts and models in hearing* (pp. 142–155). Berlin: Springer-Verlag.

Wald, G. (1945). Human vision and the spectrum. *Science*, 101, 653–658.

Wald, G. (1964). The receptors of human color vision. *Science*, 145, 1007–1017.

Wall, P.D., & Cronly-Dillon, J.R. (1960). Pain, itch and vibration. *Archives of Neurology*, 2, 365–375.

Watt, R.J. (1988). *Visual processing: Computational, psychophysical, and cognitive research.* Hove, UK: Lawrence Erlbaum Associates.

Weber, E.H. (1846). Der Tastsinn und das Gemeingefühl. In R. Wagner (Ed.), *Handwörterbuch der Physiologie* (Vol. 3, pp. 481–588). Braunschweig: Vieweg.

Wedell, C.H., & Cummings, S.B., Jr. (1938). Fatigue of the vibratory sense. *Journal of Experimental Psychology*, 22, 429–438.

Weisenberger, J.M. (1986). Sensitivity to amplitude-modulated vibrotactile signals. *Journal of the Acoustical Society of America*, 80, 1707–1715.

Wheat, H.E., & Goodwin, A.W. (2000). Tactile discrimination of gaps by slowly adapting afferents: Effects of population parameters and anisotropy in the fingerpad. *Journal of Neurophysiology*, 84, 1430–1444.

Wheat, H.E., Goodwin, A.W., & Browning, A.S. (1995). Tactile resolution: Peripheral neural mechanisms underlying the human capacity to determine positions of objects contacting the fingerpad. *Journal of Neuroscience*, 15, 5582–5595.

Zoppi, M., Voegelin, M.R., Signorini, M., & Zamponi, A. (1991). Pain threshold changes by skin vibratory stimulation in healthy subjects. *Acta Physiologica Scandinavia*, 143, 439–443.

Zwicker, E. (1974). On a psychoacoustical equivalent of tuning curves. In E. Zwicker & E. Terhardt (Eds.), *Facts and models in hearing* (pp. 132–141). Berlin: Springer-Verlag.

Zwicker, E., Flottorp, G., & Stevens, S.S. (1957). Critical band width in loudness summation. *Journal of the Acoustical Society of America*, 29, 548–557.

Zwicker, E., & Scharf, B. (1965). A model of loudness summation. *Psychological Review*, 72, 3–26.

Zwicker, E., & Schorn, K. (1978). Psychoacoustical tuning curves in audiology. *Audiology*, 17, 120–140.

Zwislocki, J.J. (1960). Theory of temporal auditory summation. *Journal of the Acoustical Society of America*, 32, 1046–1060.

Zwislocki, J.J., Ketkar, I., Cannon, M.W., & Nodar, R.H. (1974). Loudness enhancement and summation in pairs of short sound bursts. *Perception & Psychophysics*, 16, 91–100.

Zwislocki, J.J., & Sokolich, W.G. (1974). On loudness enhancement of a tone burst by a preceding tone burst. *Perception & Psychophysics*, 16, 87–90.

Author Index

A page number in **bold** refers to a figure or table.

Subject Index